Cast Iron Toy Cook Stoves & Ranges

From A to Z

Dick & Joan Ford

Schiffer Publishing Ltd

4880 Lower Valley Road, Atglen, PA 19310 USA

Disclaimer

The values provided in this book are for items in optimum original condition. Repairs, repainting, or restoration could seriously detract from the value. Values may vary widely according to rarity, condition of the piece, the location of the market, and the overall eye appeal and quality of the design and manufacturer. Prices could well vary between those at specialty antique toy shows, and those encountered at general shows and outdoor markets. A complete, correct set of cooking utensils, coal bucket and shovel, and lid lifter may also enhance the value.

All of these factors make it impossible to create an absolutely accurate price list; however, a useful general value range is stated. The values reflect what one might realistically expect to pay and are listed in U.S. dollars.

Neither the authors nor the publisher will accept responsibility for any losses or gains sustained by using this pricing (value range) guide.

The items pictured in this collectors reference guide may be covered by various copyrights, trademarks, and logotypes. Their use herein is for identification purposes only. All rights are reserved by their respective owners.

This book is for the information and use of collectors. No copyright holders authorized this book, nor furnished, or approved any of the information contained herein. This book in no way attempts to infringe on any copyright. The text and products pictured in this book are from the collection of the authors of this book and various private collectors. This book is not sponsored, endorsed, or otherwise affiliated with any of the companies whose products are represented herein. This book is derived from the authors' independent research.

Library of Congress Control Number: 2003106684

Cover and book design by: Bruce Waters
Type set in Windwor BT/Souvenir Lt BT

ISBN: 0-7643-1930-2
Printed in China
1 2 3 4

Published by Schiffer Publishing Ltd.
4880 Lower Valley Road
Atglen, PA 19310
Phone: (610) 593-1777; Fax: (610) 593-2002
E-mail: Info@schifferbooks.com
Please visit our web site catalog at
www.schifferbooks.com
We are always looking for people to write books on new and related subjects. If you have an idea for a book, please contact us at the above address.

This book may be purchased from the publisher.
Include $3.95 for shipping.
Please try your bookstore first.
You may write for a free catalog.

In Europe, Schiffer books are distributed by
Bushwood Books
6 Marksbury Avenue
Kew Gardens
Surrey TW9 4JF England
Phone: 44 (0) 20 8392 8585
Fax: 44 (0) 20 8392 9876
E-mail: Bushwd@aol.com
Free postage in the UK. Europe: air mail at cost.

Contents

Acknowledgments

Without the friendships made over our years of collecting, neither our collection, nor the enjoyment and knowledge gained in assembling it, would have been possible. The enthusiastic cooperation of good toy collecting friends, including Tiny and Connie Moyer, Ralph and Merlyn Wiesehan, Bruce and Melinda Kling, Ed and Jean Hullet, Donna and Jack Johnson, Matt Medeiros, and Ralph Hylton, has added greatly to the data and compilation of photographs for this book. We are indeed grateful for the open access to both the extensive toy catalog library and collection of our dear friends Donald and Sally Kaufman.

Indelible memories were forged with fellow collectors across the country who include toy stoves among their collecting interests, such as Don and Betty Jo Heim, Bill and Stevie Weart, Gus and Joan Majeune, and Tom and Maxine Junk.

Our sincere gratitude is further extended to "toy stover" Bruce Mumford for assistance in establishing the value guide.

A special thanks is also due to the Moyers' grandson, Gordon Moyer III, for his able assistance in photography of the Moyer collection.

Our son-in-law, Rodney Golightly, has been a sincere asset in the presentation aspects of this publication.

Foreword

"Plowing new ground" is perhaps the term which most adequately describes our efforts in writing this book on cast iron toy cook stoves and ranges. While there are already numerous books published on automotive toys, banks, and other categories of antiques, collectors and dealers needs have been devoid of a handy reference book on toy stoves and ranges. We, too, have witnessed the frustration over the years by lack of consolidated information.

The Antique Stove Association deserves special mention and applause for their publication entitled "*Antique Toy Stoves and Salesman's Samples.*" Otherwise, the study of toy stoves and ranges previously would have faced a near total uncharted wilderness.

It is our sincere hope that the publication of this new book will be a welcomed aid to the toy stove collecting hobby, inspire a new generation of enthusiasts, and serve as a useful guide to antique and toy dealers.

Introduction

Cast iron toy cook stoves and ranges are largely an American phenomenon. While there were a few exceptions produced overseas, notably in Sweden, France, and Germany, as well as Canada, cast iron toys, in general, were essentially a product of manufacturers located in the United States. However, examples produced by Skoglund et Olson, and Husqvarna, of Sweden, EH and Godin of France, and a German example made largely of cast iron are included herein. Coverage of American style cast iron stoves and ranges from "A to Z" is the primary aim of this publication. European style stoves and ranges will basically be left to an author who specializes in them.

Two categories of manufacturers were basically responsible for the toy cast iron stoves and ranges encountered by collectors and dealers, namely firms which also produced a wide selection of other toys, and firms which produced large, full-size stoves. Several of the producers of full-size kitchen stoves and ranges also produced smaller "salesman's samples", which were highly detailed scale models of the real thing, usually with an accompanying carrying case, for the salesman to actually carry from location to location to demonstrate the features to potential customers. However, the term "salesman's sample" has become highly misused and abused by applying it to toy stoves and ranges which have no connection to samples used by salesmen whatsoever.

Ladies may have started off the collecting interest in toy stoves and ranges along with their pursuit of dolls, dollhouses, kitchenware, and household miniatures in days past. At the present time many men also are keen collectors. Collecting interest in this field seems to be growing at a rapid pace, with auction prices steadily on the rise.

This book wouldn't be complete without a value guide to accompany the stove and range detailed listing. A value range will be used in most cases and it should only be used as a guide. Many toy stoves and ranges appear so infrequently on the market that pricing is based on comparative rarity. Factors affecting price include rarity, originality, extent of restoration, size and complexity, eye appeal, completeness, including loose hang-on components, condition, and correct complementary pots/pans/lid lifter.

Scope

The intended purpose of this book is to list, identify, and provide an approximate value guide for toy cast iron cook stoves and ranges produced over an approximate one hundred year period, from the 1840s to the 1940s. Toy cook stoves mirrored their full-size counterparts from approximately the mid 1840s until 1880, with the advent of the range. Ranges continued to gain in popularity, eventually superseding the earlier style cook stoves in the kitchen, and toys followed accordingly.

By the 1920s, gas ranges were replacing the older type cook stoves and ranges fired by wood and coal. Readers will note that the line of "Blue Bird" toy ranges produced by Grey Iron Casting Company of Mt. Joy, Pennsylvania during this period featured both simulated coal burners, as well as gas burners on the same range, referred to as their "2 in 1." Additionally, "3 in 1" Blue Bird toy ranges featured coal, gas, and electric burners, all on the same range. The Kenton Hardware Company featured cast iron "Universal" electric ranges in their 1931 catalog, as did Arcade Manufacturing Company with the Hotpoint. Cast iron toy production essentially ceased in early 1942, with the United States turning most manufacturing into the war effort.

This book covers cast iron toy stoves and ranges only. At the resumption of civilian goods production after the end of World War II in 1945, alternate materials were used in the production of toys, such as tin and plastic. Therefore, toy stoves and ranges in materials other than cast iron, and similar home decorative items made of pottery and ceramics, such as stove cookie jars and lamps, fall outside the intended scope of this book.

Likewise, "salesman's sample" stoves and ranges were made in exacting functional detail for demonstration purposes, and were not intended as toys, so do not fall within the intended scope of this book. These should be differentiated, and perhaps covered in a separate publication.

Heating stoves, parlor stoves, and "pot belly" stoves also form a separate toy category and are thus excluded from coverage herein. This allied category of stoves, used primarily for heating purposes, and the various other categories mentioned above are highlighted within the chapter on "Complementary Fields of Collecting."

Alphabetical Order for the Listing Arrangement "A to Z"

The term "A to Z" in the book's title has a dual meaning. While we have endeavored to be as complete and all-inclusive in our coverage of toy cast iron cook stoves and ranges as possible, we have also chosen to list them by their name, usually found on or nearby the oven door, in alphabetical order. To list them by manufacturer would be confusing to readers not already highly familiar with the subject. However, a cross-reference, by manufacturer, can be found in the Index.

A few toy cook stoves and ranges do not have a name on the oven door, removable side hearth shelf, or front area, so they will be listed under "unnamed."

Chapter One

History of Cast Iron Toy Cook Stoves and Ranges

Toys, in general, tend to mirror their real, full size counterparts, and cast iron toy stoves and ranges are no exception. Stoves started to make a serious presence in the American kitchen in the 1840s. Toy stoves date from the same period.

The "Little Eva" toy cast iron stove illustrated in the "A to Z" listing herein is a good example of an 1840s product. The removable front hearth plate states "N. S. Cate, Boston" as the manufacturer. The Boston Almanac of 1847 records N. S. Cate as a stove maker, located at 25 Sea Street. He was no longer listed in 1848, and Earl N. Cate was recorded at this same location. A later version of the "Little Eva" has been noted with the name "T. Southard" or "G. F. Filley" on the hearth plate, thus indicating the patterns for this toy stove passed from one manufacturer to another.

Progression of toy stoves into the 1850s is illustrated by the "Fairy Gift," as an example. This stove states a patent date of 1854 by R&K. "The Belle" is dated 1868, and this basic style of the cook stove stayed in production at least through the 1870s. The range was introduced about 1880, and the old style wood/coal burning cook stove was phased out. Stoves basically had black finishes, with eventual nickel trim by the 1870s. The Michigan Stove Company produced two large and impressive nickel plated toy ranges with their "Mills Range Jr." dated 1881, and the "Garland Range Jr." dated 1883. Over the next fifty to sixty years toy buyers were able to purchase the new style of nickel plated or copper flashed toy range in a dazzling choice of sizes.

By the 1920s, gas and electric ranges became available on the appliance market for the homemaker. Miniature versions were produced for children in cast iron until the outset of World War II. War production demands halted cast iron toy production and, essentially, the toy industry reverted to the use of other materials with the resumption of civilian goods production after the war. Pressure die cast and plastic materials had been making significant in-roads in toy production since the late 1930s. Reducing the cost of freight in shipping cast iron was another factor sealing the fate of cast iron toys.

Hence, present day collecting of toy cast iron stoves and ranges encompasses approximately a century of production. This span of offerings from the 1840s to the 1940s provides the stove hobby enthusiast of today with the scope to form a diverse and interesting collection.

Chapter Two

Manufacturers

The manufacturers of toy cast iron cook stoves and ranges, as stated in the introduction of this book, fall into two basic categories: toy companies which produced toy stoves and ranges along with other categories of cast iron toys, and stove and range manufacturers which produced miniature versions of their full-sized offerings. A listing of toy stove/range manufacturers follows. Some toys indicate the manufacturer, but many do not. It should be noted that often toy stoves and ranges are shown in various wholesaler or jobbers' catalogs, such as Ives, Udell & Co., etc., which were not the actual manufacturers of the item.

Toy Manufacturers of Toy Cook Stoves and Ranges

Arcade Manufacturing Company, Freeport, Illinois

An area of Freeport, Illinois called "The Arcade" was often referred to by our good friend and local fellow toy collector, the late Fred MacAdam, and was the relocation home of Arcade Manufacturing Company in 1885. Their earliest toy range is very rare, and is actually dated 1888 on the left side panel casting.

Arcade was reorganized in 1893, and their 1902 catalog shows four stoves were in production. Namely, these were the Royal, Daisy, A. C., and Gem, all in black finish. Later these stoves were painted light blue. Gas and electric ranges followed, including several private label advertising issues based on their toy Roper range, as well as Hotpoint and Peninsular.

Champion Hardware Company (The)

Champion had moved its operation from Cleveland, Ohio to Geneva, Ohio long before their two small stove offerings were produced in the early 1930s. Both the "Daisy" and the two sizes of "Geneva" are clearly marked "Champion." Toy production ceased in the mid-1930s.

Columbia Grey Iron Company

Columbia produced a "Climax Range" at their plant in Columbia, Pennsylvania in 1897. The size was 6 1/2" long, 5 1/2" wide, 4 1/2" high, with the word "Columbia" on the left side hearth shelf.

Dent Hardware Company (The)

Dent began toy production in the late 1890s, and is well respected for a quality line, including toy ranges, which continued to at least 1934. The cast iron Dent set of Chrysler Airflows and LaSalles are highly prized by automotive toy collectors, but are rare and appear to be among Dent's last efforts as there are no known catalogs for this period.

Dent produced a very nice selection of wood cooking ranges in the early 1900s, varying in size from their beautiful, deluxe 24" "The Queen" down to tiny two-burner types. Their last gasp at producing toy ranges appears to be the rare "Daisy" gas range in three sizes from 9 5/8" to 6" and the "Jean," which are illustrated in Dent's catalogue No. 10 from the late 1920s. However, of these latter gas ranges, the two smaller size "Daisy" and the "Jean" have only been encountered as brass patterns.

Undoubtedly, due to Dent's reputation for quality, they produced patterns, and possibly production quantities, for other toy companies. Detailed examination yields evidence that Hubley and Kenton were among Dent's wholesale customers. Some, but not all, of Dent's largest "The Queen" range are marked "Wing and Parsons Mfg. Co., Chicago" on the rear bolt-on plate.

The Dent factory was finally sold to other interests in the early 1970s; however, the Dent Hardware Company name still appears on the front door of the main building in Fullerton, Pennsylvania.

Grey Iron Casting Company

From their factory in Mount Joy, Pennsylvania, Grey Iron produced an impressive array of toy ranges. The company was established in 1881, and is best known for the "Crescent" and "Queen" toy ranges.

The prolific subsequent reproductions, both by Grey Iron and later manufacturers, seem to appear constantly in about every antique mall across America! While black painted reproductions of these two names are extremely common, nickel plated originals are fairly scarce.

A unique line of "Blue Bird" brand ranges included the "three in one" combination coal, gas, and electric, and "two in one" combination coal and gas offerings. Accompanying toy cast iron cookware for the larger sizes of "Blue Bird" ranges is color matched to the range in an attractive light blue/ivory combination.

In the late 1920s, Grey Iron offered several of their older style cast iron wood cooking ranges at five different price points to attract the toy buying public. For instance, the customer could choose the "Cora" range as No. 40 1/2 EB with blue enameled body, No. 40 1/2 nickel plated, No. 40 1/4 with the extension shelf but no backpiece, No. 40 with no shelves, or No. 40X with steel back and end instead of cast iron. This marketing technique was novel among the various toy stove manufacturers.

Hubley Manufacturing Company (The)

Hubley produced perhaps the highest volume and the widest selection of cast iron toys of all the U. S. cast iron toy firms, with the possible exception of Arcade Manufacturing Co. Arcade had a huge impact on the toy market with their Andy Gump car, taxi cabs, and the Ford Model T and Model A, but Hubley surpassed them all in toy cast iron ranges with their "Eagle" line. No other line offered so many sizes and types as the "Eagle" ranges, and the production volume and excellent sales distribution are still evidenced today by the frequency of Hubley "Eagles" offered to collectors.

Lancaster, Pennsylvania was the home of Hubley, and the largest toy range offered in their 1920 catalog was the 22 1/2" "Eagle." By 1925, this largest size was eliminated from production in lieu of somewhat smaller sizes. Like the other cast iron toy manufacturers, Hubley's magnificent line ceased with their turnover to wartime production in 1942.

Ideal Manufacturing Company

Located in Detroit, Michigan, Ideal Manufacturing Company developed a nice line of cast iron ranges by the 1890s. Their 1894 catalog pictures their "Ideal Range No. 5," which was their largest. This series of ranges included Ideal's No. 3, 2, 1, and 0, but no No. 4 appears to have been produced. The frequently encountered "Baby" and "Pet" ranges were also in the 1894 line-up. An increased selection of mid-size to small ranges developed into the early 1900s.

When the National Novelty Corporation was established in 1903, many toy manufacturers joined forces to form a "toy trust" to enjoy the benefits of combined production volume. Parts were made for each other between the firms in the trust. Several years later, when the "toy trust" was dissolved, some of the original firms merged or did not survive. This created much confusion for present day collectors trying to identify the correct manufacturers of a particular toy range. For example, earlier versions of the Ideal "Baby" range have the wording "Ideal Mfg. Co., Detroit Mich." below the oven door, whereas this identification is deleted on subsequent production. Likewise, the small "Perfection" range was clearly a product of Ideal before, and perhaps during, the days of the "toy trust," but after 1912, or so, this range was slightly re-designed and produced by Kenton Hardware Company.

About this same time, Ideal's cast iron ranges appear to have been discontinued, with some foundry patterns utilized subsequently by other firms. Characteristic of Ideal, and perhaps their possible predecessor or successor, is the vertical backpiece offset to the left side and the stovepipe on the right side.

Kenton Hardware Company (The)

Kenton produced locks, hardware, and eventually toys, starting in the 1890s in Kenton, Ohio. The driving force in the firm soon became Lewis S. Bixler, who had gained considerable prior experience working for his uncle, William Shimer, in Freemansburg, Pennsylvania and as a partner in the Jones & Bixler Mfg. Co. Under Bixler's leadership, Kenton rapidly expanded its toy line in the early 1900s.

Toy ranges were a mainstay in Kenton's early toy selection with the "Daisy," "Beauty," and "National," followed later by the "Model," as well made, desirable ranges to collect today. By the 1930s, Kenton surpassed even Hubley and Arcade in the diversity of types and styles of ranges produced, but many were in low volume production and, thus, are rare. Their "Royal" line of ranges was second only to Hubley's "Eagle" line in sizes and types offered. Judging by the rarity of so many of Kenton's cast iron toys, and the infrequency of them being offered today on the market, Kenton would rank third among the toy companies in sales and distribution, but certainly among the best in quality, detail, and variety.

Unlike the other cast iron toy manufacturers, Kenton resumed pouring cast iron horse-drawn and automotive toys at the change to civilian production after the end of World War II, until going out of business in 1952.

Kilgore Manufacturing Company (The)

Kilgore, located in Westerville, Ohio, was perhaps best known for its cap pistols and paper caps, and the great series of six take-apart 1933 Graham cars and trucks and other vehicular cast iron toys. Only one, rather insignificant, small 2 1/2" cooking range was produced. Listed as "No. T-7," "Gas Stove," it went through minor pattern alterations from its inception in about 1928 through the mid-1930s, and is commonly found today. This firm ceased toy production prior to the end of the Second World War.

North and Judd

For only about one year, c. 1930, North and Judd produced a limited line of mainly smaller size cast iron automotive toys, plus only one 2 3/4" gas "Anchor" range. This toy range is quite common, appears in various colors, and has a couple of minor variations, namely with and without a second door on the upper back, and has a cast anchor symbol. The anchor was North and Judd's trademark. Their location was New Britain, Connecticut, and it is mere conjecture at this point as to whether this small toy range had any connection to the full-size Anchor brand stoves and ranges produced in New Albany, Indiana. This small range is more frequently encountered than other toys in North and Judd's limited offering.

William Shimer, Son & Company

Shimer operated in Freemansburg, Pennsylvania from the mid-1870s until about 1913, and produced a nice line of cast iron ranges, varying in size from their 14 1/2" "Jupiter," down to their tiny 2 1/2", rare "36" and bank version "37." For a small company with a limited line, it is amazing that some of Shimer's larger ranges, such as the "Jupiter," "Venus," and "Mars" were produced in two very different designs. One of the two "Jupiter" designs has a striking similarity to the "Rival" range produced by J.&E. Stevens.

Shimer's nephew, Lewis S. Bixler, worked at the firm as a teenager, and later became the boss of Kenton Hardware Company.

Skoglund and Olson, A-B

This is the only non-U. S. toy firm listed in this section and it is included because this is really the only foreign cast iron toy producer which closely paralleled our domestic producers. Many U. S. automotive toy collectors include the distinctive, but limited and rare, series of cars and trucks produced by A-B Skoglund et Olson, in Gefle, Sweden as real prizes in their collections. "SetO," the abbreviated name for the company, also produced two early style cooking ranges between about 1929 and 1938, as well as a simulated tile-sided version, an oven, and a series of accompanying cast iron pots and pans.

The Swedish home appliance maker Husqvarna is known to have produced at least one cast iron toy range, which is similar to the larger "No. 2" range produced by Skoglund & Olson.

J & E Stevens Co.

Stevens was located in Cromwell, Connecticut, and was established in 1843. Subsequently, their series of "Bay State" stoves was produced in cast iron, with the sizes reduced from five to four by 1883. Casting variations on the same "Bay State" step stove design were introduced over the course of several years of production.

During the early 1900s, the "Bay State" stove series was replaced by ranges with various names and sizes. Patents were granted beginning in the 1890s, and the intricate casting details represent the height of the pattern maker's art. The largest of Stevens' cast iron ranges was the "Eclipse," at 20" in length, and it was accompanied by several other names, in descending size down to the "48" at 5" in length by 1906.

By 1924, no less than ten different types of the "Home" range were added to Stevens' offerings, as well as a small "Art" range and a two-burner version of the "48." Stevens ceased production about 1930, but left the stove collectors of today with some wonderful treasures to pursue.

Vindex Toy Div.

The Vindex line of cast iron toys was produced by the National Sewing Machine Co. of Belvidere, Illinois for only a brief period of a couple of years, c 1928-30. Only one range was offered, their cream/nile green 14" "Admiral."

Unique design nickel-plated accessories, including the lid lifter, accompanied the "Admiral." It was billed in their catalog as "A Real Toy Stove—Just Like Mother's." It is particularly difficult to locate this range 100% complete because the upper warming oven doors and main oven door handle were easily removed and lost.

Wilkins Toy Company (The)

During the mid-1890s, Wilkins was taken over by

Kingsbury, but it wasn't until 1919 that the Wilkins name was dropped in lieu of the "Kingsbury Toy Company." The business remained in Keene, New Hampshire.

Only two cast iron ranges are shown in Wilkins' 1895 catalog, the No. 201 "New England Kitchen" at 16" length and the No. 200 unmarked "Stove" at 9 1/2" length. However, the 202 range shown in the Chapter Eight listing also appears to have been made by Wilkins. These may have been Wilkins' only entries into toy cast iron stoves or ranges, and were no longer shown in their 1911 catalog.

A. C. Williams Company (The)

The present location of Ravenna, Ohio is where the company was rebuilt by Adam Clark Williams in 1893, after previous foundry fires. Sad irons (heavy flat irons with a curved raised arch for a handle, used for pressing clothes) and still banks, plus hardware items, were the mainstay of the business until the toy line was expanded to include cast iron "wheel toys" in the 1920s.

"Dime stores" served as a high volume outlet for the inexpensive Williams toy line, which explains why their small upright cast iron "Royal" ranges are so commonly found today. These gas ranges, catalogued as No. 67 for the one burner and No. 167 for the two burners, and the slightly larger No. 267 with "Broiler," completed Williams stove offerings until the updated green/red "Williams" 3 5/8 model was added about 1936.

The latest observed Williams toy catalog price supplement is dated April 1, 1937 and by 1938 the firm had ceased toy production, but remained in the hardware manufacturing business.

Stove and Range Companies which Produced Miniature Versions of Their Full-Size Offerings

Several firms which produced full-size stoves and ranges also produced toy versions. These tended to be large and heavy, and were more likely sold by the same appliance and hardware dealers who handled the large cast iron stoves and ranges.

This category of toy offerings is included in the alphabetical "A to Z" listing, and cross-referenced in the Index. Some of them, quite possibly, could have doubled as samples used by traveling salesmen to demonstrate the full-size version, as opposed to stoves and ranges made by the toy companies in the prior section. However, we have endeavored to include only cast iron toy stoves and ranges in this publication and, thus, exclude obvious salesman's samples with their accompanying carrying cases.

Chapter Three

Terminology

Every collecting field seems to develop its own terminology and toy stoves and ranges are no exception. A few should be highlighted.

Stove vs. Range

It is well to clarify the difference between a "stove" and a "range." We all tend to use the term "stove" as generic and all-encompassing, and inclusive of ranges and wood/coal/gas and electric fueled heaters, which is a carryover from the earliest days of separate kitchen cooking and room/home heating devices in the 1800s.

This book centers initially on cast iron toy cook stoves which began appearing as replicas of the real size kitchen stoves in approximately the 1840s. The hearth is located on the shorter front side of the stove and the stove pipe is at the rear, with the oven doors opening from either of the longer sides.

Cast Iron Toy Stove

While the early type toy cook stoves are largely cast iron and very heavy, the later toy ranges often only have the top cooking surface, base, removable side hearth, extension shelves, and removable backpiece actually made of cast iron. The sides, bottom, and inner dividers of ranges are often just formed from tin plate to reduce weight and shipping cost. Some toy range manufacturers, such as J&E Stevens, did actually utilize intricately detailed castings on all four sides. Smaller size toy ranges, by some manufacturers, do incorporate cast iron panels on all six sides.

In summary, when we refer to "cast iron" toy stoves and ranges, tin components both inside and outside are often utilized to reduce weight and shipping cost, including the removable stovepipe.

Cast Iron Toy Range

The "range" appeared in the kitchen starting about 1880, and gradually replaced the "stove" in most domestic applications. Wood and coal were utilized for fuel in the kitchen range well into the 1930s and later. By the 1930s, however, gas and electric ranges were rapidly replacing the older type wood fueled cooking ranges. A "range" is decidedly different in appearance from a "stove." A range has a detachable hearth shelf at the left side, roughly half-way up the height of the range. Also, a range has a central single oven door, often accompanied by an upper warming oven incorporated into an ornate high backpiece. The stovepipe may be concealed within the backpiece. On ranges without a central high backpiece, the stove pipe is found centered, or offset to the right side. Ranges often include a removable extension shelf on the right side of the cooking surface. Hence, just because many of the toy ranges have the stovepipe offset to the right, when facing the oven, that is not grounds to call it a "stove."

Toy Stoves and Ranges vs. Salesmen's Samples

All too often the term "salesman's sample" is used to incorrectly add desirability to a toy cast iron stove or range. This misused term is widely seen describing an actual toy at toy shows, antique malls, and over the Internet. Salesman's sample stoves and ranges normally include a carrying case, with handle, and are produced as a miniature version, in exacting detail, of actual full-size stoves for demonstration purposes to prospective customers.

It should be noted that some of the larger toy stoves/ranges, however, could have also served as a salesman's sample, such as the Monarch, St. Clair, Great Majestic Jr., etc.

Step Stoves

The term "step stove" is used to describe the earlier version of stoves with the lower "step" front hearth serving as an integral portion of the stove. "Bay State" toy stoves by J&E Stevens serve to illustrate the hearth area and the cooking surface as two separate steps. Stoves with a third "step" are also encountered.

Go-Withs

"Go-withs" are advertisements, accessories not initially sold with a toy stove or range, signs, etc. which can be added to further accent a display or provide additional interest and knowledge. Several examples are shown in a separate chapter on this topic.

Lid Lifter

A lid lifter is the handle or tool used to lift the burner cover or lid.

Chapter Four

Reproductions and Their Detection

No collecting field seems to escape the menace of reproductions, and toy stoves and ranges are no exception. Fortunately, the toy stove and range reproductions made to date are relatively few in type, not of the earlier pre-1880 designs, and are fairly easy to recognize. The original toy stoves and ranges were either enameled, electro oxidized, or nickel plated. They often were made with tin sides and back panels, but at least the doors, trim and trivets were nickel plated. The reproduced copies are distinguished as follows:

1. Rougher cast surfaces, not smooth to the touch like the originals.

2. They are invariably painted black with the door name possibly a lighter color.

3. They lack detail, like the omitted wording on the left side of the "Dolly's Favorite," or the missing back panel on the "Novelty."

4. They often utilize Phillips™ head screws and bolts instead of the slotted head type originally used.

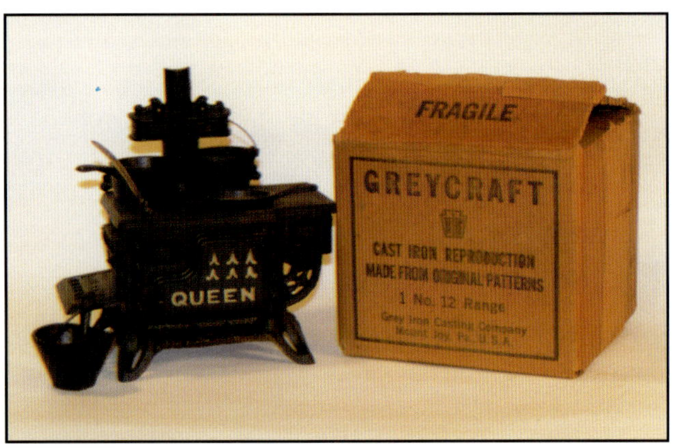

While this "Queen" range with its original box is an official reproduction by the original manufacturer, the foundry patterns were later used by other superseding firms. These later reproductions with a black finish are extremely common.

Keeping in mind that the original cast iron stoves and ranges are a minimum of sixty years old, some degree of toning and light surface rust should normally be readily noted on originals. Reproductions started appearing in the 1950s and the "Queen" and "Crescent" originally produced by the Grey Iron Casting Co. of Mount Joy, Pennsylvania, were by far the most prolific. As can be seen in the above photo, Grey Iron actually identified their initial reproductions of "Queen" and "Crescent" ranges on their boxes. These were sold under the "Greycraft" name, and the boxes stated, "Cast Iron Reproduction Made From Original Patterns." The foundry patterns for these two toy ranges were transferred to at least one other foundry in the U. S., and then the latest repros are marked "Taiwan."

Besides the "Queen" and "Crescent," reproductions are often encountered of the "Dolly's Favorite," "Novelty," and "Dot" ranges. Reproductions of Kenton's 1936 gas "Table Top Stove" in green/nickel are often seen. Some of these reproductions are currently sold in gift shops and all can be found on the Internet. Buyer beware!

Recent Creations in Cast Iron

In addition to reproductions of cast iron ranges, toy enthusiasts will encounter new post-war brands on the market. Commonly encountered names recently produced in cast iron are: "Bonnie-Glo," "American," in two sizes, "Eager Beaver," made in Japan, "Princess," (a smaller version on Grey Iron's "Queen"), the Canadian "Belanger Centenaire 1867-1967" with "Royal" on the oven door, "Boyd's," "Busy Bee," "Dot," with a coin slot and "Cook with Cash" imprinted inside the oven door, "National Journal," "Iron is Livitamin" (re-do of the Grey Iron "Queen"), "Ranch Craft," and "Reno, Nevada 1868."

Reproduced Loose and Hang-On Parts for Toy Stoves and Ranges

There are several active sources which will repair toy stoves and ranges and reproduce missing parts. This is a mixed blessing. Simple repairs to original stove components, such as replacing a broken tab on a side shelf to make it once again functional, could be welcomed by most collectors. Restoration of the total stove, re-plating, and replacement of missing parts could significantly detract from the value. This practice, and its resultant effect on value, could be compared to the difference in value and appeal between a restored versus an original Buddy L pressed steel truck. One only has to check recent toy auction results to see the high premiums collectors place on the originality factor.

Chapter Five

Additional "Go-Withs" for Toy Stoves and Ranges

The term "go-withs" is aptly applied by collectors of a certain primary series of collectibles to describe related accent items which enhance the display. Advertising items from firms which produced full-size stoves and ranges often serve as attractive "go-withs" for similar toy stoves and ranges.

As an example, the toy "Loth's Fuel Saver" range made by Arcade Manufacturing Company is shown ably accented by a Loth's advertising blotter and various cast iron ashtrays and miniature pots. There are a number of "go-withs" which add nice accents to the "Dolly's Favorite" toy range made by the Favorite Stove and Range Company in Piqua, Ohio. These could include miniature frying pans used to advertise both Favorite and their various stove dealers, mid-size Favorite frying pans, a tape measure, a pinback button, etc. Framed magazine advertisements of stoves and ranges and trade signs act as interesting "go-withs" for their small toy versions.

Stove and range company "go-withs" which are used to accent toy versions can form a very nice subcollection in their own right, which can be seen in the following illustrations:

GW1. The Michigan Stove Co. A customized 1920s pressed steel delivery van is a great "go-with" for any stove and range collection! *Courtesy of Tiny Moyer.*

GW2. The Michigan Stove Co. This facsimile of a Mills Range in the form of a paperweight is an excellent "go-with" for an 1881 toy Mills Range Jr. range. *Courtesy of Tiny Moyer.*

GW3. Arcade Mfg. Co.'s toy "Loth Fuel Saver" range is shown accompanied by various Loth advertising items.

GW4. O K Stove and Range Co. Opening stove top advertisement of 3 1/4" length, in cast iron, for O K stoves and ranges. The accompanying toy O K range was made by Ideal Mfg. Co. in Detroit, Michigan.

GW5. Opening stove top advertisement of 3 1/4" length in cast iron for "O K Stoves and Ranges." This same basic stove advertisement was used by several stove/range companies. *Photo courtesy of Donna Johnson.*

GW6. Grey Iron Casting Co. This firm used the "Greycraft" name for their 1950s vintage reproduction "Crescent" and "Queen" black finished ranges. Boxed sets of pots and pans were likewise offered. *Courtesy of Tiny Moyer.*

GW7. Quite a variety of cast iron animals were used as advertisements by the stove and range companies.

GW8. Abendroth Bros., New York. Cast iron frogs were a favorite advertising item for some stove and range companies. *Courtesy of Tiny Moyer.*

GW9. Mt. Penn Stove Works. Several different small tea kettles were made by stove and range firms, and were identified as advertisements to nicely accompany toy stoves and ranges. *Courtesy of Tiny Moyer.*

GW10. Pocket mirrors were an excellent advertisement method utilized by many stove and range companies.

GW11. An additional selection of stove and range advertising pocket mirrors. *Courtesy of Tiny Moyer.*

GW12. Cribben & Sexton Company. "Tip Trays" were issued by various stove and range companies for advertising purposes. *Courtesy of Tiny Moyer.*

GW13. Hubley had the exclusive right to produce these two c. 1929 toy cast iron electric refrigerators (No. 837 and No. 838). They can be used to accompany a display of toy ranges, such as Hubley's "Eagle" line.

GW14. Dent Hardware Co. Brass factory patterns for Dent's large size lid lifter and burner cover.

GW15. Bussey & McLeod Stove Co., Chicago. The "Gold Coin" toy range is displayed with this large porcelain sign. *Courtesy of Tiny Moyer.*

GW16. Majestic Manufacturing Co. This large sign serves as a nice backdrop to display the toy Great Majestic ranges. (The author is caught photographing in the reflection!) *Courtesy of Tiny Moyer.*

GW17. The Michigan Stove Co. The firm's logo on this large porcelain sign for their "Garland" line parallels the logo on the toy 1883 "Garland Range Jr." range. *Courtesy of Tiny Moyer.*

GW18. Rathbone, Sard, & Co. This colorful porcelain sign shows the firm's logo and can nicely accompany the toy Acorn range. *Courtesy of Tiny Moyer.*

GW19. Phillips & Buttorff Mfg. Co. The toy "Model Enterprise" stove can be nicely displayed with this tin sign. *Courtesy of Tiny Moyer.*

GW20. Peninsular Stove Co. This nice porcelain sign shows the Peninsular logo used by their factory in Detroit, Michigan, and can serve as a fine back drop for the toy Peninsular range by Arcade Mfg. Co. *Courtesy of Tiny Moyer.*

GW21. Engman-Matthews Range Co., Goshen, Indiana. A nice sign to accompany the toy "Range Eternal." *Courtesy of Tiny Moyer.*

GW22. The Prizer-Painter Stove Works, Reading, PA. Although Prizer is not known to have produced any toy stoves, their advertising items are attractive "go-withs" in a collection. *Courtesy of Tiny Moyer.*

GW23. Numerous home decoration items in the form of stoves and ranges, such as this "Acme" die cast table lamp, cookie jars, etc. can be used to expand a collection theme.

Chapter Six

Complementary Fields of Collecting

Once a collector delves into the fascinating field of toy cast iron stoves and ranges, the door soon opens up to other allied and complementary fields of collecting. Collectors can expand into various categories:

CF1. An original company sign and recipe book enhance the display of a toy size "Range Eternal."

Salesman's Samples

Although salesman's samples are generally readily discernable from toy stoves and ranges as true-to-scale functional miniatures, and are normally accompanied by their carrying case, they can form a separate adjunct collection. Some collectors of salesman's samples further expand to incorporate other desired sample categories, such as other types of kitchen appliances and washing machines.

CF1A. This Bridge-Beach "Superior" is an excellent example of a true salesman's sample, complete with carrying case. The exquisite detail was necessary for the salesman to demonstrate the features of the real thing. *Courtesy of Tiny Moyer.*

CF2. Salesman's sample heating stoves exhibit the same working detail, in miniature, as cook stoves. Here is a very nice example showing the Simmons Wilson 316.

Other Cast Iron Toy Kitchen Items

Toy cast iron kitchen cabinets, furniture, refrigerators, food grinders, and ice shavers all form a nice allied collection. Arcade Manufacturing Company of Freeport, Illinois produced a large dollhouse, plus a smaller version thereof, from the late 1920s to the early 1930s, which would accommodate the various rooms of toy cast iron furniture they also produced. Both an electrical kitchen set and a gas kitchen set were available for the dollhouses.

The Hubley Manufacturing Company of Lancaster, Pennsylvania offered a selection of cast iron Alaska brand ice boxes and G. E. refrigerators to complement their extensive offerings of Eagle brand ranges. The Alaska ice boxes came complete with plates of artificial food and glass cubes of artificial ice. The G. E. electric refrigerator toys included cast iron ice cube trays.

CF4. Early style toy refrigerators serve to complement toy ranges of the period. Both sizes of G. E. refrigerators were made by Hubley Manufacturing Co., and the larger, double-door type is very scarce.

CF3. The Hubley "Eagle" range in the center is unusual in red. It has its correct, original pot and pan and is nicely accented by an accompanying G. E. refrigerator and kitchen "Cabinette," also in red. This trio forms an attractive display.

CF5. Hubley Manufacturing Co. produced Alaska brand toy ice boxes in several sizes, which came with glass ice cubes and plates of artificial food. The loose nameplate is from a full-size Alaska, and was made by the Dent Hardware Co.

Toy Stoves and Ranges Made of Materials Other Than Cast Iron

A huge and diverse selection of toy tin stoves and ranges was produced both in the United States and overseas, in varying sizes. Some were electrified. Pressure die cast stoves were also produced, as well as those made of plastic, ceramic, pottery, rubber, and wood.

Home decorators recently (1960s-1970s) sold a die cast replica of a toy range called "West" in a copper flashed finish, with non-removable burners. "Stern Industries" is clearly marked on the back.

CF7. A 4 1/8" plastic range is a decorative addition to a collection, or a kitchen. This colorful item holds salt and pepper shakers in its burners, and is shown with its original box.

CF6. This fired ceramic "Dot" range serves to hold the accompanying salt and pepper shakers, in the form of a tea kettle and coffee pot.

CF8. A wide range of toy tin ranges is available, many with electrical heating elements. Many were produced outside North America. European versions often incorporate cast iron cooking surfaces and trim in their basically tin construction.

CF9. This German toy range with tall, removable stovepipe is basically constructed of tin plate, with nice brass trim and a cast iron cooking surface. (Stovepipe is shown in front of stove, as it is too tall to be pictured on stove.)

Toy Scales

A collection of cast iron toy scales acts as an interesting supplement to a collection of cast iron toy stoves and ranges.

CF10. Numerous toy scales and their accompanying weights and weighing pans were produced by toy companies. Shown are three examples, with their original boxes, made by Keystone, A. C. Williams, and Kenton.

Toy Heating, Parlor, and "Pot Belly" Stoves

A closely parallel series to collect encompasses toy stoves which were primarily utilized for heating, instead of cooking. These were produced in the same wide range of materials as cook stoves and ranges. Toys and banks appear in the form of heating, parlor, and "pot belly" stoves in a considerable array of types and sizes.

CF11. The right hand version of this toy "pot belly" stove is an advertisement for Sears, and the one on the left was a sales award in 1968.

CF10A. "The Pet" is a nice example of a toy parlor stove, made by Adams, Pickover & Co., with a patent date of 1857. Another close version of this parlor stove was made by John Clark of Albany, New York.

CF12. Toy heating stoves and furnaces can take the form of a still bank, such as the "Glow-Boy" on the left, made by the Globe Stove and Range Co., of Kokomo, Indiana. The Godin "Radiolette," on the right, is a French example of a parlor stove.

CF12A. This still bank states "SAVE YOUR MONEY AND BUY A GAS STOVE."

CF14. Abendroth Bros. of New York produced this "Gem" heating stove still bank. *Courtesy of Tiny Moyer*

CF13. The Detroit Stove Works produced this miniature "Jewel" parlor stove. *Courtesy of Tiny Moyer*

CF15. A two-burner portable gas stove such as this could have served various purposes. This toy version is 8 3/4", and is marked "R Y" underneath.

Miniature Cast Iron Stove Advertising Frying Pans

Many stove manufacturers and dealers utilized small size frying pans with lettering either inside, outside, or both, to advertise various brands of stoves and ranges. Most are cast iron, but occasionally tin examples are also encountered.

CF16A. Advertising small frying pans are as diverse as stove companies themselves! Canadian firms and dealers issued them, as well as U. S. stove/range producers and dealers.

CF16. Miniature stove and range advertising frying pans can be hung on hooks and framed to form a separate interesting collection.

Paper Advertising and Stove Company Letterheads, Envelopes, and Billheads

Magazine ads and paperwork used by stove companies act as interesting and valuable information about the firm's products, which the toy stoves and ranges represent. For example, the accompanying company postcard displayed with the "A-B" toy range, made by Arcade Manufacturing Co., readily informs the toy owner or viewer that the A-B Stove Company was located in Battle Creek, Michigan, and was in business in 1914.

CF19. There are numerous collectibles to accompany an 1883 "Garland Range Jr.," by The Michigan Stove Company, but possibly none more colorful than this advertisement.

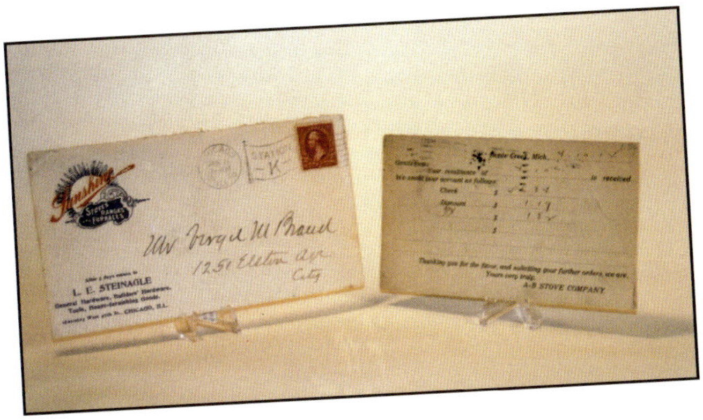

CF17. Displays of a toy "Sunshine" range by Orr Painter & Co. of Reading, PA., and an "A-B" range by the A-B Stove Company of Battle Creek, Michigan can be nicely enhanced by the addition of these contemporary paper items.

CF18. These two paper items can add to the display appeal of two popular toy ranges made by the Arcade Mfg. Co., the "Peninsular" and the "Quick Meal."

CF20. Although The Favorite Stove and Range Co. of Piqua, Ohio only produced one toy range, the "Dolly's Favorite," a wide variety of advertising items for this firm can be assembled, including paper goods.

Chapter Seven

Identification and Variations

Who Made What, and When?

Hopefully, this publication will serve to aid the toy stove/range enthusiast with quick reference as to which manufacturer made which item. In most cases, the range name is on the oven door, whereas the name of an early stove can generally be noted on the front hearth plate, which is often removable, or on the body of the stove above the hearth.

Once the name of the stove/range has been determined, that particular item can easily be located in the "A to Z Alphabetical Listing and Price Guide" chapter of this publication. Stoves/ranges without a name thereon are listed under "Unnamed." Every toy stove and range known to us at the time of publication has been included in the listing. The "Index" will serve to cross-reference all the toy stoves and ranges made by each particular manufacturer. A few stoves/ranges are listed but simply cannot be assigned to a particular manufacturer at this point in time, so these can be found in the "Uncertain" area of the "Index," under the letter "U."

ID1A. The lady's vignette on the oven door, raised within a circle, and the unusual scalloped base create an identification challenge.

ID1. The number "202" on the oven door of this 8" range relates to Wilkins' catalog number sequence, and follows the two ranges in their 1895 catalog, i.e. the No. "200" and No. "201."

Fortunately, most of the early toy stoves have the manufacturer's name cast prominently thereon, in addition to a stove name. Also, most of these manufacturers also produced full-size stoves. If this identifying information, however, is shown on the removable front hearth plate, and the plate has become lost and missing from the stove over the years, only visual comparison with another complete version of the stove would yield the necessary identifying information for the collector.

Likewise, most of the initial toy ranges, introduced about 1880, were also identified as to manufacturers, such as the "M.I.M." in 1881 and the "Garland Range Jr." in 1883, both by The Michigan Stove Co., and the "Jewel Range Jr." by the Detroit Stove Works. Progressively, toy manufacturers began to produce the overwhelming majority of toy ranges to add to their line of cast iron horse-drawn and automotive toys. Toy wholesalers and department stores were clamoring for a fuller line of cast iron toys to fulfill the ever increasing public demand. Few toy companies identified their products, as did Arcade Manufacturing Company, Kenton Hardware Company and, to some extent, Ideal Manufacturing Co. The majority did not, such as Hubley, Stevens, Grey Iron, Shimer, Dent, and Williams.

Only by thoroughly researching toy company catalogs and wholesalers' catalogs, and by assembling old toy advertisements from such magazines as "Playthings," could a beginning collector ascertain which toy manufacturers produced which toy stoves/ranges. Hopefully, "The A to Z Alphabetical Listing and Price Guide" chapter will relieve the reader from this burdensome task.

Nevertheless, many toy stoves and ranges are not represented in any known toy catalogs, such as the bulk of Dent Hardware Company's and Ideal Manufacturing Company's products. Efforts to unearth literature in the Fullerton, Pennsylvania area, where Dent was located, and in the Detroit, Michigan area, where Ideal was located, have proven to be fruitless, appertaining to their toy ranges. Only by years of examining other collections, asking limitless questions to seasoned collectors, and carefully comparing the similarities in basic designs and pattern making, plus such subtleties as the commonality of burner lids and lid lifters found on original toy ranges, and the casting similarities of frame base skirting designs, does a reasonable degree of certainty develop in assigning manufacturers' names.

From such analysis, it appears Ideal Manufacturing Company in Detroit was the only firm which characteristically produced toy ranges with the rear, vertical, removable backpiece offset to the left (sometimes incorporating the loose tin stovepipe into the backpiece design, and sometimes merely placing the tin stovepipe to stand alone separately, in the right rear area). This common Ideal characteristic is noted on the large "Ideal No. 5" range, as well as their "No. 3, 2, and 1," and carries over into their versions of the "Perfection," "Pet," small "Grand Jewel," and others. Ideal joined the infamous "toy trust," comprised of a large number of individual toy manufacturers in the early 1900s, and when the "trust" was dissolved after a few years it appears Ideal had ceased production and their patterns were modified and utilized by other manufacturers, particularly Kenton Hardware Co. The small "Perfection" range, for example, now had the loose, vertical backpiece or stovepipe centered at the rear of the cooking surface, rather than offset to the left.

It also appears that certain toy manufacturers produced detail parts for other toy firms and, in some cases, complete toy ranges. Dent Hardware Company produced at least one size "Eagle" gas range for Hubley, which is identical to Dent's "Daisy." The basic design, as well as the three large, centralized "beads" in the lower base frame skirting area, indicates with considerable certainty that Dent produced at least the patterns for Kenton's "Rival" stove, if not the complete stove, on behalf of Kenton, which Kenton illustrated in their catalog and marketed. These two examples are shown in the accompanying photos. The design commonality is readily apparent. Dent utilized this same design frame base on several of their ranges, such as their "Queen" of this same overall size and their "Waldorf."

ID2. The early "Perfection" range on the left has the tell-tale offset rear backpiece and ornate base design, which point this range in the identification direction of Ideal Mfg. Co. When Kenton Hardware Co. subsequently produced the "Perfection" in various types and sizes, the backpiece/stovepipe was centered and the base design was simplified.

ID3. Gas range shown with factory pattern "Daisy" oven door, plus same basic range modified for Hubley Mfg. Co. by Dent with the "Eagle" name. The "Eagle" brand modification is an example of casting pattern work performed by Dent Hardware Co. for Hubley Mfg. Co.

Size Identification

In general, the height and width (depth) dimensions of a toy stove or range are a function of the length dimension. Also, the total height is often difficult to state accurately because of varying lengths of stovepipes furnished over a period of several years of production, and because often the same toy range was marketed both with and without a fancy vertical backpiece, to allow for varying price levels of the same toy. Hence, in identifying the size of an early style stove the overall length from front to back will be the one dimension stated in the listings, and from side to side, including removable side hearth and extension shelves, for ranges.

Size dimensions stated in catalogs will generally have been taken from foundry patterns and, hence, are 1/8" to 1/4" larger than actual specimens due to the shrinkage factor after the castings are poured and cooled in actual production.

Design Variations

Some toy stoves and ranges had short production lives, for a variety of reasons, and are rare and only known in one version. However, even some of the lower volume manufacturers, such as Shimer Toy Company, managed to produce at least two readily discernable major design variations of many of their range offerings over a short time period. Examples are Shimer's "Jupiter" and "Venus" ranges. Other ranges, such as Hubley's "Eagle" and Kenton's "Royal" were produced in relatively large quantities and over a period of several years. Minor design modifications of the same basic range were made over the years of production and, unless truly significant, will not be highlighted. For example, a tin side panel could have been substituted for an earlier cast iron panel to achieve cost savings and weight reduction.

ID4. The "Rival" range is shown in Kenton's catalog, whereas the "Queen" range is a Dent Hardware Co. product. The base design and other features are common between these two ranges, as well as other Dent offerings, indicating Dent produced patterns, and possibly complete ranges, for other toy companies.

Confusing Effects of the Early 1900s Toy Industry Trust

During the first few years of the 1900s, many of the toy manufacturers progressively joined forces with allied firms to form an industry trust known as The National Novelty Corporation and, subsequently, The Hardware and Woodenware Company. The umbrella effect of the trust provided increased production volume for the individual firms, and many detail parts and components were manufactured by one firm for another. Patterns and tooling were transferred between firms, and some firms in the trust either "died" or were absorbed by other firms, such as Shimer, Ideal, and Jones & Bixler. The trust was dissolved approximately a decade later with some firms, such as Kenton, rising victorious from it. Casting patterns were modified for commonality to suit the production techniques of the succeeding firm. As mentioned earlier, for example, a range like the small "Perfection" started as a typical Ideal Manufacturing Co. design with a left side offset, vertical backpiece, before the formation of the toy trust, and emerged after the dissolution of the trust as a Kenton product, modified with a central backpiece. Shimer's line of toy ranges disappeared completely.

Needless to say, the tumultuous effects of the toy trust add considerable confusion in attempts to accurately identify the makers of many of the early twentieth century ranges.

Correct, Matching Pots, Pans, and Accessories for a Toy Stove or Range

As a general rule, the various toy firms made sets of pots, pans, and coal buckets and shovels, etc. to either accompany their offerings of stoves and ranges or to be bought separately. Some of the earlier stove producers did so also, while some did not, expecting the toy customer to buy a set of "Wagner Ware" or "Griswold" small pots and pans to add play value to the toy stove. Only through the information provided by original toy catalogs and advertisements can we ascertain with any degree of reliability which accessories go with which stoves and ranges, as their size is often interchangeable.

Kenton sold a larger, separate set of four accessories, consisting of a teakettle, handle pot, stew pan, and frying pan, and also included it to complement their three largest ranges, the "Model," "Globe," and "Star." Kenton's smaller ranges had a corresponding smaller set of accessories included with each toy range. The larger four accessories are clearly marked "Kenton Brand" underneath. The smaller size accessories are not marked, and neither are most of those produced by the other toy companies.

Hubley did cast a part number into their pots and pans for a certain period of years, but this practice was not consistent. Likewise, G. F. Filley of St. Louis gave his stove accessories a "G" number, which accompanied the "Charter Oak" stoves. Beyond these examples, few others are encountered with identifying markings.

"O. P. & C." markings are seen on accessories issued with the toy "Dainty" and "Sunshine" ranges by Orr, Painter and Co., Reading Stove Works, Reading, PA. Where possible, the photos accompanying the "A to Z Alphabetical Listing and Price Guide" chapter include, to the best of our knowledge, those stove/range top accessories which are correct for that particular item. Missing accessories were not available at the time of photography. Generic accessories are shown in some cases on stoves and ranges not known to have been originally issued with complementing accessories, or in photos noted from other collections.

Lid Lifters

Each of the early stove companies which made toy stoves generally included a lid lifter of their own design. Wherever possible, the correct lid lifter is shown on the stove. The toy companies, likewise, each produced a lid lifter to accompany each of their toy coal/wood range offerings. Although largely interchangeable, they are decidedly discernable by manufacturer, as shown below:

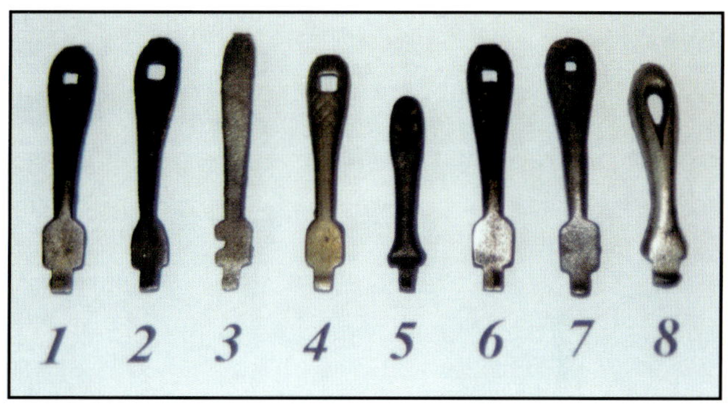

ID5. Small size lifters included on ranges by toy companies: 1-Arcade, 2-Dent, 3-Grey Iron, 4-Hubley, 5-Ideal, 6-Kenton/Shimer, 7-Stevens, 8-Vindex. Ideal also used a lifter similar to Kenton's. *Note:* Hole size and grid/no grid pattern can vary.

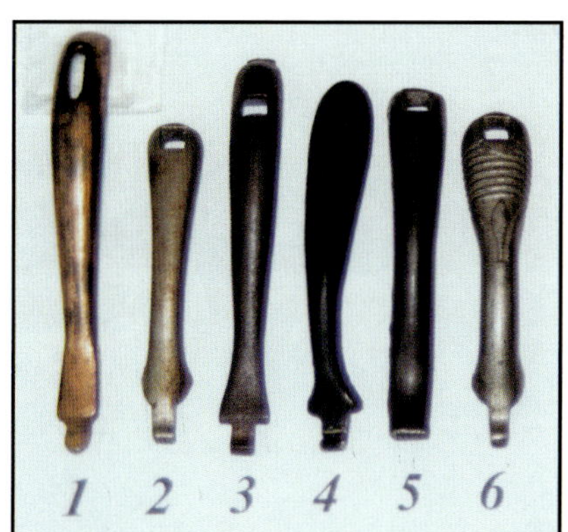

ID6. Larger size lid lifters included on ranges by toy companies: 1-Dent, 2-Grey Iron/Kenton, 3-Hubley, 4 & 5-Ideal, 6-Stevens. *Note:* Arcade , Shimer, and Vindex utilized only the smaller size lid lifters.

Chapter Eight

A to Z Alphabetical Listing and Price Guide

Optimum values stated are reflective of complete, original items with no missing parts, in generally above average condition. Correct, original lid lifters, pots/pans and accessories also add to the value. Hence, a value range is stated.

When considering the minimum value of any small, four-burner range, for example, the burner lids are worth $15-$20 each, at least; the tin stovepipe is worth $20 or more, and any removable side shelves are worth at least $30 each, for original parts that can be used to complete an incomplete range. Hence, the bottom value for any given complete, small range, just for the parts value, has to be $75-$100. A removable rear backpiece, or more elaborate side shelf, would add further value.

A

001. A.1: Range made with left hearth shelf only. A later, similar design range was made by Shimer, and subsequently Kenton. 7 3/4" l., mfgr. Ideal. $125-175.

002. A.1: Range made with left hearth shelf only. Similar design later made by Kenton. Earlier type with circular wreath on door was made by Ideal. 7 3/4" l., mfgr. Shimer. $125-175.

003. A.1: Range made with left hearth shelf only; marked "Kenton Brand" on back of range. Note: Name "A.1." on hearth shelf, and fancier base design. Earlier type with circular wreath on door was made by Shimer. 7 3/4" l., mfgr. Kenton. $125-175.

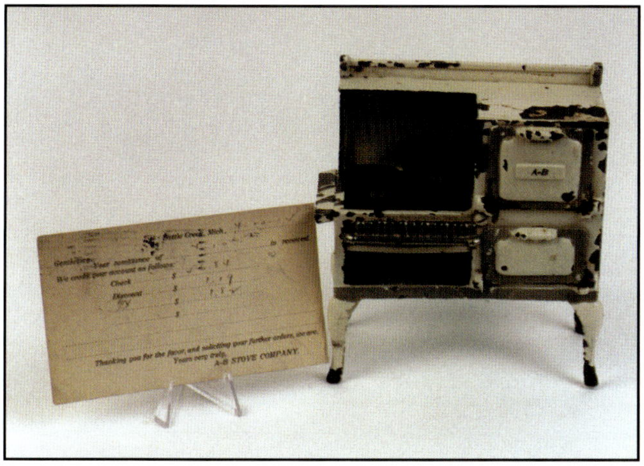

004. A-B: Gas range with private label, similar to Arcade's "Roper" and other private label ranges by Arcade. Photo shows 1914 correspondence from A-B Stove Co., Battle Creek, MI. 5 1/2" l., mfgr. Arcade. $300.-450.

005. ABC No. 3: Cook stove with porcelain and nickel finish. 17 1/2" l., mfgr. ABC. $1,750.-2,250. *Courtesy of Tiny Moyer.*

006. ABC No. 3: Cook stove with porcelain and nickel finish. Different view of above listing, shown with salesman's sample ABC washing machine. 17 1/2" l., mfgr. ABC. $1750-2250. *Courtesy of Tiny Moyer.*

007. AC: Small range produced without side shelves. 5" l., mfgr. Arcade. $100-150.

008. Acme: Range produced in elaborate casting detail with two side shelves. Patented Sept. 23, 1902. 14 3/4" l., mfgr. J&E Stevens. $500-750.

009. Acme: Range produced with left hearth shelf only; nickel/black finish. Cat. no. 935 1/2. The unusual curved tin stovepipe includes damper and is held by the fancy, nickel plated, vertical backpiece. "Acme" was also available nickel plated. 13" l., mfgr. Kenton. $500-650.

010. Acme: Nickel plated range produced with left hearth shelf only and ornate removable backpiece. Cat. no. 935. 13" l., mfgr. Kenton. $400-500.

011. Acorn: Larger, elaborately cast range, with left hearth shelf and right reservoir. Very rare. Additional "go-withs" issued by the mfgr. as advertising give-aways highlight this range. 19 1/2" l., mfgr. Rathbone, Sard & Co. $5,000-6,000. *Courtesy of Tiny Moyer.*

012. Admiral: The only range produced by Vindex; nickel plating with Nile green and cream trim. Supplied with unique set of nickel plated lid lifter, coal scuttle & shovel, stew pan and skillet. Vindex toys were only produced for a year or so, c. 1929, by National Sewing Machine Company, Belvidere , IL. 14" l., mfgr. Vindex. $1500-2000.

013. Adrian: Range with left side hearth shelf; available in nickel plated or copper flashed finish. Noted only in this one size. 10" l., mfgr. Dent. $250-350.

014 (not shown). American Jewel: No details available at time of printing.

015 (not shown). American: Range with black finish, noted in two sizes. The initials "A.T.F." are on the oven door. Both have two side shelves and a backpiece with stovepipe. The larger size "American" resembles the "Crescent" range of this size by Grey Iron Casting Co., except for the oven door. The smaller size also appears to have been cast from Grey Iron patterns. 13 1/2" & 9 1/2" l., mfgr. possibly Grey Iron or a successor. $200-300. & $75-100.

016 (not shown). Amy: Nickel plated range with four burners and the stovepipe offset to the right side rear. This range was possibly made by Ideal Mfg. Co., Detroit, MI and was shown in an early Lord & Taylor catalog.

017. Anchor: Small gas range with single upper warming oven. North & Judd produced a limited line of cast iron toys, c. 1930, including this one gas range in two varieties, with only one upper door and with two upper doors. 2 3/4" l., mfgr. North & Judd. $50.-100. *Courtesy of Tiny Moyer.*

018. Anchor: Small gas range with two upper warming ovens. North and Judd produced a limited line of cast iron toys, c. 1930, including this one gas range in two varieties, with only one upper door and with two upper doors. 2 3/4" l., mfgr. North & Judd. $50-100. *Photo courtesy of Matt Medeiros.*

019. Anita: This range is unusually large for a child's toy, but since it came from a toy museum it is hereby listed. 30" l., mfgr. Union Stove Works, New York. $2,500-3,500.

020. (not shown) Annie: Early black finish stove, with four burners and removable hearth plate with the date Dec. 25, 1878. The length and design are like the "Little Willie" stove shown herein, by Abbott & Noble, except that the name "Annie" is shown on the removable front hearth plate, rather than above the front opening doors. 14" l., mfgr. Abbott & Noble. Philadelphia. $500.-750.

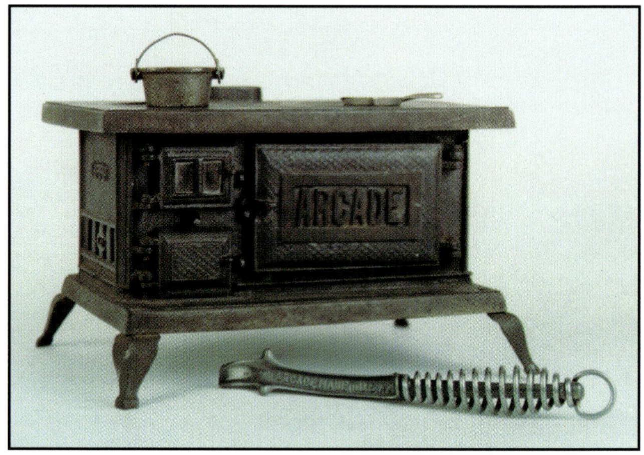

021. Arcade: This range has the date 1888 cast on the left side, which is early for Arcade. Arcade produced this range in 1888, in Freeport, IL., prior to moving to their new foundry facilities in Feb. 1893. 11" l., mfgr. Arcade. $1,500-2000.

022. Ark: Small range produced without side shelves. Kenton produced only one variety of the Ark range. 5" l., mfgr. Kenton. $100-150.

023. Art: One of Stevens' smaller ranges, which was produced in one size without side shelves. 5" l., mfgr. J&E Stevens. $125-150.

B

024. Baby: Black finished range marked: "Ideal Mfg'r. Co. Detroit, Mich." The 1894 catalog shows an accompanying set of teakettle and three pot/pans. Other examples of this range have been noted which are unmarked as to mfgr. and are nickel plated. A copper flash version has been noted with a backpiece, and was probably produced later by Kenton. 15 1/2" l., mfgr. Ideal. $600-800.

024A. Baby Grand: Impressive range with black finish, white porcelain back containing upper shelf, and three nickel plated front doors. 12" l., mfgr. Charles Fawcett, Sackville, N.B., Canada. $1,250-1,750.

025. Baby: This tiny range is the smallest of Hubley's wide selection, No. 853. The removable backpiece design is the same as Grey Iron's "Tot," but attaches differently. The brass/lead pattern for this range was included with other brass patterns from the Dent Hardware Co. *Note:* The front door panel design is the same as a series of other ranges of this small size, namely the "Cute," "Dot," "Pet," "Gem" (sold by Arcade), "Home," and "Nit'" (sold by Shimer), so which firm made which details, possibly during the toy trust days of the early 1900s, remains a mystery. 3 1/2" l., mfgr. Hubley. $50-100.

026. Baby: (pattern) This brass/lead pattern was located among other brass patterns from Dent Hardware Co. Hubley modified the manner of attaching the removable backpiece from "twist" to "slide in." See comments for above listed item. 3" l., mfgr. possibly Dent , Shimer, or Grey Iron for Hubley. $250.

027. Baby: See above two listings. Hubley sold this small range with just a tin stovepipe, or with a cast iron removable back piece. No. 849. 3" l., mfgr. Hubley. $50-100.

028 (not shown). Baby Hazel: Black finish range, with four nickel plated legs and oven door frame. The oven door center insert states "Baby Hazel" in blue letters on white porcelain. *Note:* This range is similar in design to the "Wetter," "U. S.," "Mascot," and "Hanks Junior" ranges, which were produced by the U. S. Stove Corp. of South Pittsburg, Tennessee, formerly H. Wetter Mfg. Co./Bluff City Stove Works of Memphis, Tennessee, and the Mascot Stove and Mfg. Co., of Chattanooga, Tennessee. 17" l., mfgr. Rome Stove & Range Co., Rome, GA. $3,500-4,500.

029 (not shown). Bailey : (Boyd's), Modern ranges with black finish, c. 1997. The larger of these two home decorator/gift shop ranges stated "Boyd's Bears & Friends" on the oven door, with a paw print. The box has the name "Bailey's Cook Stove." 13" l., & 8" l., mfgr. Boyd's Collection Ltd. (The). Made in China. $10-15.

030. Bardes: Gas range similar to Arcade's "Roper" and "Peninsular," private labeled for "Bardes." 5 1/2" l., mfgr. Arcade. $300-450. *Courtesy of Tiny Moyer.*

031. Bay State or Eureka: This Stevens stove is similar to the "Eureka" listed, except for the sliding hearth plate. 10 1/2" l., mfgr. J&E Stevens. $500-750. *Courtesy of Tiny Moyer.*

032. Bay State: Four sizes of "Bay State" stoves are shown. The 1883 Stevens catalog lists only four sizes, but a 5th (11" l.) size is listed elsewhere, which may be actually the Stevens "Eureka" stove listed herein. Variations of the "Bay State" stoves have two closing front hearth doors, paint trim differences, etc. 4", 6", 7", 10" l., mfgr. J& E Stevens. $150-450.

033. Bay State: Variation in design can be readily noted on these two 4" examples of the Stevens "Bay State" series. 4" l., mfgr. J&E Stevens. $150-175.

034. (not shown) Beauty: Large, elaborate nickel plated range. This range is dated 1891, and has a separate base. Although the basic range is nearly identical to the "Garland Range Jr." and the "M.I.M." produced by the Michigan Stove Co. which are dated 1883 and 1881, respectively, the backpiece and rear upper shelf are totally different with the letter "B" on either side of the stovepipe. 20" l., mfgr. The Michigan Stove Co. $1750-2500.

035. Beauty: An electro-oxidized range, as shown in Kenton's 1900 catalog. A left side hearth shelf was provided, but no extension shelf. 16" l., mfgr. Kenton. $600-800.

036. Beauty: A nickel plated porcelain range, c. 1900. A left side hearth shelf was provided, but no extension shelf. 16" l., mfgr. Kenton. $700-900. *Courtesy of Tiny Moyer.*

036A (not shown). Beauty: Range with black finish, two trivets on center backpiece, and the name "Beauty" within an oval on the oven door. 16" l., mfgr. uncertain. $400-600.

037. Belle (The): A black finished stove, dated 1868. The burner lids are distinctive. Like so many of the early stoves, this toy stove could have been produced by a manufacturer of full size stoves of the period. 10 1/8" l., mfgr. uncertain. $650-850.

038. Belle: Nicely detailed range, produced with left side hearth shelf, but no extension shelf. 7 3/8" l., mfgr. Dent. $300-400.

039. Bess: 1890s style range, produced with left side hearth shelf but not an extension shelf. This range exhibits design characteristics of both "Ideal" and "Kenton" of the late 1890s period. See the "Bird" range, as well as the 1894 "Pet" range by Ideal and the larger 1900 "Beauty" range by Kenton. 10" l., mfgr. Kenton or Ideal. $300-450.

040. Bird: Copper flashed range produced with detailed base and left side hearth shelf. This range appears in an 1899 jobber's catalog and exhibits design characteristics of both "Ideal" and "Kenton" of that period. See the 1894 "Pet" range by "Ideal," and the larger 1900 "Beauty" range by "Kenton." 11" l., mfgr. Kenton or Ideal. $300-450.

041. Bird: Nickel plated range, produced with plain base and left side hearth shelf. Similar to above listing. This range appears in an 1899 jobber's catalog and exhibits design characteristics of both "Ideal" and "Kenton" of that period. See the 1894 "Pet" range by "Ideal," and the larger 1900 "Beauty" range by Kenton. 10 1/2" l., mfgr. Kenton or Ideal. $300-450.

042. Blue Bird: Largest size combination "3 in 1" coal, gas, and electric range. Blue/white/nickel No. 170 series. This range was sold with and without "electrical equipment" to illuminate two of the burners, giving an imitation fire appearance. 11 3/8" l., mfgr. Grey Iron. $300-500. *Courtesy of Tiny Moyer.*

043. Bluebird: Second largest combination "3 in 1" coal, gas, and electric range; cream and blue colors, shown with correct three piece pot/pans, plus lid lifter. This range, and the larger size, were sold with and without "electrical equipment" to illuminate two of the burners, giving an imitation fire appearance. No. 160 series. 10 1/2" l., mfgr. Grey Iron. $300-500.

044. Blue Bird: Combination "2 in 1" coal and gas range, with back shelf instead of warming ovens. No. 142 W. This "2in1" range was also made in 9" and 7" lengths (Cat. No. 152 & 132). 8" l., mfgr. Grey Iron. $200-300.

045. Blue Bird: Combination “2 in 1” coal and gas range; cream color with blue trimmings, No. 120W. This “2 in 1” range was also made in 9", 8", 7", and 4 1/4" lengths. (Cat. Nos. 150, 140, 130, 120). 5 1/2" l., mfgr. Grey Iron. $175-200.

047. Blue Bird: Combination “2 in 1” coal and gas range, with single warming oven. 4 1/2" l., mfgr. Grey Iron. $125-175.

046. Blue Bird: Combination “2 in 1” coal and gas range; blue color with cream trimmings. No. 120. 5 1/2" l., mfgr. Grey Iron. $175-250.

048. Blue Bird: Gas range, No. 220. 6 1/2" l., mfgr. Grey Iron. $75-100.

049 (not shown). Blue Bird: Small gas range with upper back shelf. No. 210. 4" l., mfgr. Grey Iron. $50-75.

50. Bonnie Glo: Modern, c. 1980s range marked "JW." 4" l., mfgr. John Wright Inc., Wrightsville, PA. $10-15. *Courtesy of Tiny Moyer*

052. Britt & Folger: Cook stove with front hearth plate, size n/a., mfgr. Britt & Folger. $800-1000. *Photo courtesy of Ralph and Merlyn Wiesehan.*

051. Braddick Range No. 9: An early black finished range with a distinctive appearance. 9 1/2" l., mfgr. Braddick. $800-1000. *Courtesy of Tiny Moyer.*

053. Buck's, D: Cookstove, marked "D. Buck's Patent," 11" l., mfgr. D. Buck. $600-800. *Photo courtesy of Ralph & Merlyn Wiesehan.*

055. Buck's Brilliant: c. 1889 range. Note close similarities with Michigan Stove Works "M.I.M." and "Garland Range, Jr.," dated 1881 and 1883, respectively. 20 1/4" l., mfgr. Buck's Stove & Range Co., St. Louis, MO. $2000-2500. *Photo courtesy of Ralph & Merlyn Wiesehan.*

054. Buck's Sterling Junior, No. 100: Upright design and rarest of the Buck's series. 13 1/2" l., mfgr. Buck's Stove and Range Co., St. Louis, MO. $2500-3500. *Photo courtesy of Ralph and Merlyn Wiesehan.*

057. Buck's Junior 2: Cast iron range with porcelain and nickel finish. 22" l., mfgr. Buck's Stove and Range Co., St. Louis, MO. $2,250-2,850. *Courtesy of Tiny Moyer.*

056. Buck's Junior: Cast iron range c. 1889. Note close similarities with Michigan Stove Works "M.I.M." and "Garland Range Jr." dated 1881 and 1883, respectively. 20" l., mfgr. Buck's Stove & Range Co., St. Louis, MO. $1500-2000. *Courtesy of Tiny Moyer.*

058. Buck's Junior 3: Cast iron range with left side hearth shelf. 18 1/2" l., mfgr. Buck's Stove and Range Co. St. Louis, MO. $1,800-2,500. *Courtesy of Tiny Moyer.*

059. Buck's Junior 4: Cast iron range with tin sides. 15 1/4" l., mfgr. Buck's Stove and Range Co., St. Louis, MO. $1,800-2,400.

060 (not shown). Busy Bee: Modern small range with black finish and non-removable burner lids. 4" l., mfgr. John Wright, Inc., Wrightsville, PA., and later "made in Taiwan." $10-15.

C

061. Carrie: Four burner stove. 13 1/2" l., mfgr. G. F. Filley, St. Louis, MO. $500-700. *Courtesy of Tiny Moyer.*

062. Carrie: Mfgr. unmarked, but same design and size as the "Lilly," "Julia," and "Little Fanny" stoves made by the Philadelphia Stove Works/Orr Painter & Co. 11 1/4" l., $500-750. *Courtesy of Tiny Moyer.*

063. Champion Hot Blast: This ornate range is so detailed that it could have served as a salesman's sample rather than a toy. This range was patented June 21, 1892. 10 1/4" l., mfgr. Champion Steel Range Co., Cleveland, Ohio. $3,500-4,500. *Courtesy of Tiny Moyer.*

064 (not shown). Champion: Range with copper flash finish, ornate backpiece and bust of a lady within a large circular wreath on the oven door. This range has been reported, but not verified. The ornate backpiece is like one reported on a "Home" range by Ideal of similar size. 15 1/2" l., mfgr. Ideal. $600-800.

065. (not shown) Charlotte: Stove with four burners and removable hearth and legs; similar to the "Little Eva." Foundry patterns for this basic stove appear to have been originally used by N. S. Cate of Boston. Progressively, the same basic design was used by Southard, Robertson & Co., N.Y.C. 10 1/4" l., mfgr. probably T. Southard, Robertson & Co., N.Y.C. $500-750.

067. Charm: Small range with fluted border around cast iron base and stove top. 5" l., mfgr. Grey Iron. $100-125.

066. Charm: Small range with ornamental, removable backpiece. Backpiece and fluted design on horizontal front borders are common to Grey Iron's "Home" range. 5" l., mfgr. Grey Iron. $125-175.

068. Charm: Small range with plain border around cast iron base and stove top. This "Charm" range has an unusual pattern design in the tin sides and back. 5" l., mfgr. Grey Iron. $100-125.

069. Charter Oak No. 30: This model of the Charter Oak stove contains a high reservoir with removable lids. Dated 1871 and marked "Filley's Pat'. Hot Air Flue," "Pat'. Extended June 14, 1867." 24 1/2" l., mfgr. G.F. Filley, St. Louis (Excelsior Stove Works). $1,500-2,400.

070. Charter Oak No. 503: This model of the Charter Oak stove contains a reservoir and closet and, like the No. 103, was marked "Filley's Pat'. Hot Air Flue," "Pat'. Extended June 14, 1867." 24 3/4" l., mfgr. G. F. Filley, St. Louis (Excelsior Stove Works). $1,500-1,800. *Courtesy of Tiny Moyer.*

071. Charter Oak No. 103: Stove with removable legs, marked "Pat. 1881," "Filley's Pat'. Hot Air Flue," "Pat'. Extended June 14, 1867." *Note:* Pots & pans are marked on bottom with a "G" number. The teakettle lid advertises "G.F. Filley, St. Louis" An earlier issue of this stove is dated 1876. Also, a version of the no. 103 was produced with two hearth shelves, one above the other. G. F. Filley of St. Louis, Mo. also used the company name of Excelsior Mfg. Co., 19 3/4" l., mfgr. G.F. Filley, St. Louis (Excelsior Stove Works). $900-1,300.

072. Choice: Model No. 50 1/2 nickel plated range; also sold as No. 50 1/2 EB, enameled body, No. 50 1/4, nickeled with extension shelf only, and No. 50, nickeled without shelves. 10 1/2" l., mfgr. Grey Iron. $200-300.

071A. Charter Oak No. 103: Stove with removable legs. Another view of the above item, showing the nickeled door pattern, but same markings. 19 3/4" l., mfgr. G. F. Filley, St. Louis (Excelsior Stove Works). $800-1,200. *Courtesy of Tiny Moyer.*

073. Clara: Nickel plated range with tin sides and no side shelves. 5 1/2" l., mfgr. uncertain. $250-350.

075. Columbian Stove Works: This four-burner cook stove was located in S. Australia, as is. Perhaps it once contained a loose hearth cover plate. The design is similar to the "Little Eva," "My Favorite," and "Merry Christmas," perhaps produced by T. Southard or G. F, Filley. 13 1/2" l., mfgr. Smith & Wellstood. $600-800.

074 (not shown). Cleveland: Die cast modern bank range dated 1980 on the bottom; painted black and silver with tulip designs on the oven door, base and backpiece. 7 1/2" l., mfgr. Mattel. $15-25.

076 (not shown). Columbian: Stove with left side, full length side hearth shelf; appears identical to the "Royal American" stove, also sold by Bridgeford, except for the name on the oven door. 18 1/4" l., mfgr. Bridgeford & Co., Louisville, Ky. $1,750-2,250.

077 (not shown). Columbia: Range with four burners, and right side rear stovepipe. The name "Columbia" is on the left side hearth shelf. The oven door merely has a slight design pattern. This is the only range appearing in Columbia's 1897 catalog (No. 17). 6 1/2" l., mfgr. Columbia Grey Iron Co., Columbia, PA. $500-750.

078. Comet: Nickel plated range, produced with removable hearth shelf only. 8" l., mfgr. Ideal. $125-175.

079. Cora: Range with enameled body and two side shelves. Catalog No. 40 1/2 EB. This range was also produced as No. 40 1/2 (nickel plated with two side shelves), No. 40 1/4 (with stovepipe and hearth shelf) and No. 40 (with no shelves), and No. 40X with steel back and end instead of cast iron. 9" l., mfgr. Grey Iron. $350-450.

080. Cora: Range with nickel plating and two side shelves, shown in photo with original box. Catalog No. 40 1/2. This range was also produced as No. 40 1/2 EB (blue enameled body), No. 40 1/4 (with stove pipe and hearth shelf) and No. 40, (with no side shelves). 9" l., mfgr. Grey Iron. $175-250.

081. Cora: Range with nickel plating and hearth shelf only. Catalog No. 40 1/4. This range was also produced as: No. 40 1/2 EB (blue enameled body), No 40 1/2 (nickel plated), both with two side shelves, and No. 40 (with no side shelves). 7" l., mfgr. Grey Iron. $125-175.

082. Cotton Plant: Cook stove with sliding hearth plate. 11 1/2" l., mfgr. Abendroth Bros., N.Y. $1,500-2,000.

084. Crescent: Copper finished range, original, and the largest size; produced with two shelves and stovepipe only(no backpiece). This range was located in Canada. Reproductions with black finish and backpiece are still being produced currently. 13" l., mfgr. Grey Iron. $600-750.

083. Crescent: Nickel plated range; original, and the largest size. Cat. No. 18. The two shelves and backpiece are removable. Reproductions with black finish are still being produced currently. 13" l., mfgr. Grey Iron. $600-750.

085. Crescent: Mid-size range with blue enameled body and nickel finish. Reproductions with black finish are still being produced currently. 9 3/4" l., mfgr. Grey Iron. $300-350. *Courtesy of Tiny Moyer.*

086 (not shown). Crescent: 1950s reproduction range and original box; black finish. Grey Iron Casting Co. utilized the original foundry patterns for this reproduction range and marketed this item and their similar "Queen" shown herein under their "Greycraft" brand. Boxes marked "No. 20" were used for both. *Note:* Some of Grey Iron's patterns were later, c. 1970s, used by John Wright Foundry in Wrightville, PA., such as the 13" "Crescent" range patterns. 9 3/4" l., mfgr. Grey Iron. $100.-150.

087. Crescent: Nickel plated range, original, and the smallest size "Crescent." produced with the hearth shelf only. Reproductions of the Crescent ranges are still being produced currently with a black finish. 71/2" l., mfgr. Grey Iron. $175-250.

089. Cute: Small nickel plated range with all cast sides and removable backpiece. This is one of a family of such small ranges with integrally cast hearth shelf and the underneath turnkey take-apart feature. Names on the oven door are "Pet," "Home," "Dot," "Gem" (sold by Arcade), and Shimer's "Nit'" made without the hearth shelf. 3 1/4" l., mfgr. Shimer. $85-125.

088. Crown: Range with blue enamel and nickel finish. The catalog No. is 100 1/2 EB. This range was also supplied with nickel finish, and also as a version with only the hearth side shelf. 13 1/2" l., mfgr. Grey Iron. $500-750.

D

090. Dainty (c. 1880s-90s): Black & nickel range equipped with right extension shelf only. Pan is marked "O.P. & C.," offered in Montgomery Ward's 1889 and 1894 catalogs. 12 1/2" l., mfgr. Orr. Painter & Co.(Reading Stove Works, Reading, PA.) .$600-800.

092. Daisy: Larger range; photo example is missing its two side shelves. Kenton's 1900 catalog shows another style backpiece. 17 1/2" l., mfgr. Kenton. $700-900.

091. Daisy (c. 1900): Larger size range. Kenton Hardware Mfg. Co. used two different vertical decorative backpiece designs. 17 1/2" l., mfgr. Kenton. $800-1,000. *Photo courtesy of Ralph and Merlyn Wiesehan.*

093. Daisy: Gas range shown with factory pattern for the oven door. Largest of three sizes. This is the only size noted as a production example. The two smaller sizes appear to be patterns only. 9 1/2" l., mfgr. Dent. $450-500.

095. Daisy: Factory pattern gas range by The Dent Hardware Co.; mid-size of three sizes. No production example of this size has been noted. 7 1/2" l., mfgr. Dent. $650-750.

094. Daisy: Gas range shown with factory pattern "Daisy" oven door, plus same basic range modified for The Hubley Mfg. Co. by Dent with the "Eagle" name. The "Eagle" brand modification is an example of casting pattern work performed by The Dent Hardware Co. for The Hubley Mfg. Co. 9 1/2" l., mfgr. Dent. $450-500, & $750-1,000.

096. Daisy: Factory pattern gas range by The Dent Hardware Co.; smallest of three sizes. No production example of this size has been noted. 6 1/4" l., mfgr. Dent. $600-700.

097. Daisy: Small brass pattern for a "Daisy" range. No production example of this range has been noted. 5" l., mfgr. Dent. $500-600.

098. Daisy: Range with black finish and shown with original box and accessories. 6" l., mfgr. Arcade. $250-350.

099. Daisy: Small range No 545. 2 7/8" l., mfgr. Champion. $25-50.

100. Delight: Nickel plated range with hearth shelf. Except for the name on the oven door, this range is similar to the copper flashed "Bird" range listed. Both exhibit design characteristics of both "Kenton" and "Ideal" ranges of the c. 1899 period. 9 1/2" l., mfgr. Kenton or Ideal. $300-450.

101. Dictator Jr.: Black finish cook stove, dated 1870. 11" l., mfgr. Detroit Stove Works. $1,000-1,200.

102. Doll (The): Black finish cook stove. The letters "G" and "W" are on opposite sides of the hearth. It is noted that "G" & "W" was in business as late as Jan. 10, 1973, when they opened a modern new foundry in Landisville, PA. 13" l., mfgr. Rikeman & Seymour-Peekskill. $1,000-1,200.

103. Doll (The): Black finish cook stove. This example does not contain the hearth markings found on the previous listing of "The Doll". 13" l., mfgr. uncertain. $1,000-1,200. *Courtesy of Tiny Moyer.*

105. Dot: Nickel plated range with removable backpiece and original pot and pan. This range was sold with and without the pot and pan shown, and also with a lacquer finish. 3 1/2" l., mfgr. Grey Iron. $100-150.

104. Dolly's Favorite: Large porcelain finish range. This is the only toy range produced by "Favorite" in the 1920s, and is seen with a blue or gray porcelain and nickel finish or a black finish. 18 1/2" l., mfgr. The Favorite Stove & Range Co. Piqua, Ohio. $1,500-2,000.

107 (not shown). Dot: Small nickel plated range, with stovepipe, and circular wreath design on the oven door containing the range's name, like Shimer's "XX" small range. 5" l., mfgr. Shimer. $150-200. *Photo courtesy of Donna Johnson.*

106. Dot: Nickel plated range with removable backpiece. This same small range is shown in both the Kenton catalog (No. 900/901) and the Grey Iron Catalog (No. 24) of the 1920s. The Kenton version is marked "Kenton" on the bottom, with a hole pattern of four groups of nine holes each. A later reproduction bank version of this range, with a black finish, has the wording "Cook With Cash" below the coin slot when the oven door is opened. 3 5/8" l., mfgr. Kenton and Grey Iron. $50-75.

108. Dude: Cookstove. 19" l., mfgr. Central City Stove Co. $1,200-1,500. *Photo courtesy of Ralph and Merlyn Wiesehan.*

107A. Dot: Small nickel plated range still bank, with all cast sides and stovepipe. This is one of a family of such small ranges with integrally cast hearth shelf and underneath turnkey take-apart feature. Names on the oven door are "Cute," "Pet," "Home," and "Gem" (sold by Arcade). 3 1/4" l., mfgr. Shimer. $85-125.

E

109. Eager Beaver: C. 1960s range with hibachi type cooking surface. This item is just marked "Made in Japan." 5 3/4" l., mfgr. uncertain. $25-50.

110 (not shown). Eagle: Coal Ranges: approximately 26 different sizes and variations of "Eagle" named ranges, which Hubley called "Toy Models of the Coal Burning Variety," were offered in the period between Hubley's c. 1920 and 1929 catalogs. The name "Eagle" was spelled out in varying designs on the oven door, and was also shown on the removable left side hearth shelf. Essentially all types were available with a choice of either an electro-oxidized or nickel plated finish in 1920, but only nickel plating was offered by 1929.

A few smaller sizes offered early had only a right side stovepipe, whereas the reduced number of "Eagle" offerings by 1929 all contained a separate rear vertical backpieces enclosing a central stovepipe. By 1929, the larger lengths of "Eagle" coal ranges had been eliminated, from 22 1/4" down to 16". By 1932, only six sizes were available, with 11 1/2" being the largest. No toy ranges of any type were offered in the 1935 Hubley catalog.

"Eagle" gas ranges, 12" & 10" sizes of coal ranges with a bust of an eagle on the oven door, a small 4 1/2" size, two-burner range with a spread eagle design, rather than wording on the oven door and, finally, two small size ranges with "Star" and "Baby" on the oven door round out Hubley's total line-up of toy ranges. These are all covered separately, by name, in the alphabetical listing. Sizes: 22 1/4" down to 4 3/4" l., mfgr. Hubley. $1,800 down to $100.

111. Eagle: Largest of Hubley's "Eagle" ranges, No. 879. This size "Eagle" range was available in the No. 1 catalog c. 1920, but not in the No. 2 catalog dated 1929. Most sizes were available in either nickel plated or electro-oxidized finish. 22 1/4" l., mfgr. Hubley. $1,250-1,800.

113. Eagle: Electro-oxidized finish range, No. 863C. 11" l., mfgr. Hubley. $300-350. *Courtesy of Bruce and Melinda Kling.*

114. Eagle: Nickel plated range, No. 887. 9 1/2" l., mfgr. Hubley. $200-250.

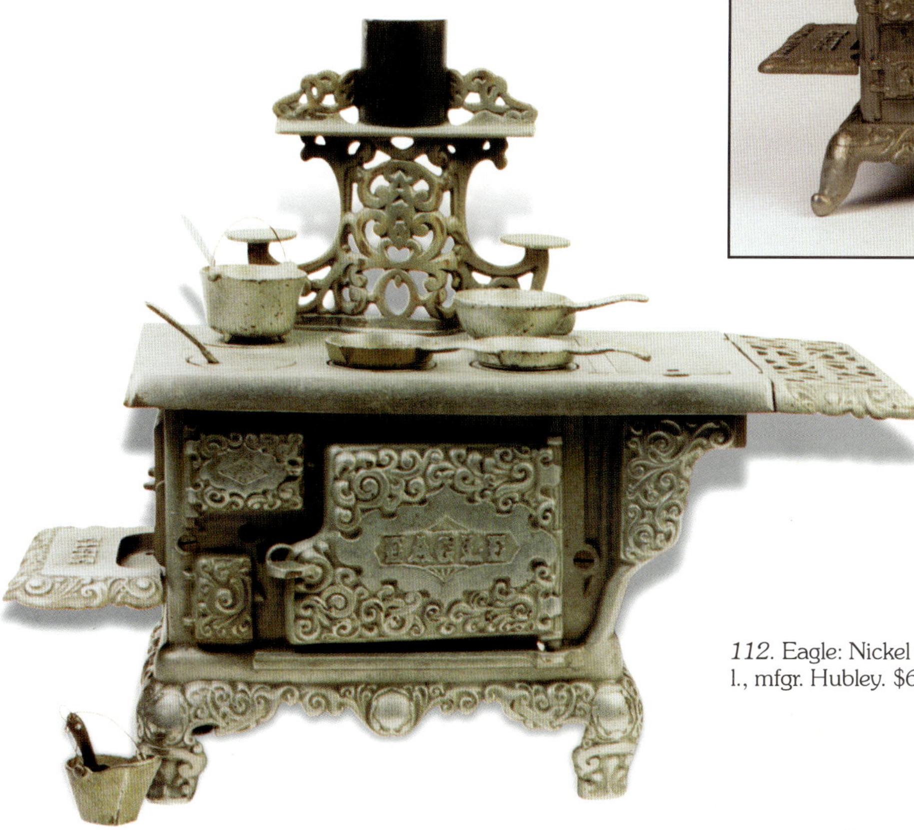

112. Eagle: Nickel plated range, No. 892. 18 1/2" l., mfgr. Hubley. $600-800.

115. Eagle: Nickel plated range, No. 895, shown with original box. 6" l., mfgr. Hubley. $150-200.

117. Eagle: Nickel plated range, produced with no side shelves and stove-pipe. No Hubley catalog reference noted, but similar to No. 852. This "Eagle" range was found with a group of patterns from The Dent Hardware Co. 4 1/2" l., mfgr. Hubley. $100-125.

116. Eagle: Nickel plated range, No. 857. 6 3/4" l., mfgr. Hubley. $125-150. *Courtesy of Tiny Moyer.*

118. *Eagle*: Nickel plated range with an eagle's bust on oven door. No. 860, except produced without right side extension shelf, which would add another 2" total length. 10" l., mfgr. Hubley. $250-300.

119. Eagle: Nickel plated range with spread eagle design on oven door. No. 838, has two burners. This small range was also offered with just a rear stovepipe (No. 889) and possibly a four-burner variety, (No. 883). 4 1/2" l., mfgr. Hubley. $75-125.

120 (not shown). Eagle: Gas Ranges: Hubley's gas ranges were patented in 1926-27. They were offered with three oven compartments on the larger sizes, as well as the patented imitation flame feature, by 1932, on five of the sizes. In total, eight styles and sizes of "Eagle" gas ranges were offered, in colors of white, blue, or green. An example of the 6 1/4" size in red is also shown herein. 11 1/2" down to 4 1/4" lengths, mfgr. Hubley. $600-75.

121. Eagle: Gas range, No. 898 G. This is the largest of several sizes of gas ranges by Hubley. 11 1/2" l., mfgr. Hubley. $500-600. *Courtesy of Tiny Moyer.*

123. Eagle: Gas range with imitation gas flame feature. No. 864G. 8 1/2" l., mfgr. Hubley. $250-300.

122. Eagle: Gas range, No. 896. 9 3/4" l., mfgr. Hubley. $300-350.

124. Eagle: Gas range with imitation gas flame feature. No. 872G. 6 1/4" l., mfgr. Hubley. $150-200. *Courtesy of Bruce and Melinda Kling.*

126. Eagle: Gas range, No. 871. 4 3/4" l., mfgr. Hubley. $75-100.

125. Eagle: Gas range No. 872, with two correct, original utensils. This range is the centerpiece of an attractive grouping of three red Hubley kitchen toys. 6 1/4" l., mfgr. Hubley. $250-300.

127. Eagle: Selection of gas ranges of various sizes. Mfgr. Hubley. $600-75. *Courtesy of Tiny Moyer.*

128. Eagle: Pattern gas range made by The Dent Hardware Co. for The Hubley Mfg. Co. This pattern range was located with a group of Dent Hardware Co. patterns and is indicative of the pattern work Dent performed. 10" l., mfgr. Dent. $500.

129. Eclipse: Early stove with black finish, four burners and four legs. Name is above the oven door. 11" l., mfgr. uncertain. $800-1,000. *Photo Courtesy of Tiny Moyer.*

130. Eclipse: Largest size of the Stevens line of well engraved ranges. The "Eclipse" was produced from c. 1890s up until at least 1924. This correct ornate teakettle was only supplied on the "Eclipse" and "Rival" ranges. The "Eclipse" range was also produced without the right hand extension shelf. 19 1/2" l., mfgr. J&E Stevens. $2,500-3,000.

131. Eclipse Box: Original dovetail wooden box for the Stevens "Eclipse" range. Mfgr. J&E Stevens. $200-250. *Courtesy of Tiny Moyer.*

132. Elektrik: Gas range, 9 1/2" l., mfgr. Dent. $600-750. *Photo courtesy of Ralph and Merlyn Wiesehan.*

133 (not shown). Etna: No further details available at time of printing. Mfgr. Orr, Painter & Co., Reading, PA.

134 (not shown). Eureka: Cook stove. Although the mfgr. is not indicated on the accompanying box, this step stove and its burner covers are similar to Stevens' "Bay State" stove series. Tin broiler is original. 10.5" l., mfgr. J&E Stevens. $1500-2,000. *Photo courtesy of Matt Medeiros.*

135. Eva: Copper flash range, No. 928. The elaborate version of the "Eva" range was a new offering in Kenton's 1905 catalog, but only a scaled down version was offered by the early 1920s, with no extension shelf and only a rear stovepipe. This version was then re-introduced in Kenton's early 1930s catalogs. As with most Kenton wood cooking ranges, finishes were offered in nickel plating, black/nickel finish, and electro oxidized. 11 1/2" l., mfgr. Kenton. $300-350.

F

136. Fairy Gift: Black finish stove, marked "pat'd. 1854". 6 3/4" l., mfgr. R&K. $750-1,000.

137. Fairy Queen: Black finish stove. 17 1/4" l., mfgr. uncertain. $750-1,000. *Courtesy of Tiny Moyer.*

138. Favorite: Range with green and nickel finish and removable back panel with two warming ovens. A new item in Kenton's 1931 catalog, the "Favorite" range was produced in three sizes: 8 3/4", 7 5/8" and 5 3/4". Size shown, 8 3/4" l., mfgr. Kenton. $500-600.

139 (not shown). Fortune Junior: Range with left side hearth shelf, and name of mfgr. on front hearth. c. 1890. 11 5/8" l., mfgr. Thomas, Roberts, Stevenson & Co., Philadelphia. $1500-2000.

140. Forty-Eight (48): Four-burner range shown with original box. 5" l., mfgr. J&E Stevens. $125-175. *Courtesy of Bruce and Melinda Kling.*

141. Forty Eight (48): Two-burner range. 4 3/4" l., mfgr. J&E Stevens. $125-175.

G

142. Garland Range Jr.: Large, elaborate, nickel plated range. This range is dated 1883, and has a separate base. When the oven door is opened, a movable scrollwork shelf lowers, and again raises upward when the door is closed. 20" l., mfgr. The Michigan Stove Company. $2,500-3,000.

143. Garland: Cast iron and tin range with high back shelf. Although The Michigan Stove Co. produced the full-size Garland ranges, it is not confirmed if this company also produced this type of "Garland" toy range. 9" l., mfgr. uncertain. $450-500.

144 (not shown). Gem, (The): Early black finish stove, with maker's name on hearth plate, and removable legs. This same style stove is shown herein, with the name "Little Eva" on the removable hearth plate by N. S. Cate, Boston. Other manufacturers' names associated with the subsequent use of the foundry patterns to produce this basic stove with different names are: "Scranton Stove Co.," "T. Southard," and "G. F. Filley." John Clark also produced a toy parlor stove marked "The Pet" and "John Clark, Albany" on the hearth. A similar design to Clark's "The Pet" parlor stove, shown in the chapter entitled "Complementary Fields of Collecting," is marked "Adam, Pickover & Co." with a patent date of 1857. 14" l., mfgr. J. W. Clark, Albany. $500-750.

145. Gem: Nicely detailed cast iron range, shown complete with correct accessories. 8" l., mfgr. J&E Stevens. $250-350.

146. Gem: Small range with all cast iron panels and backpiece. Although this small range is shown in the Arcade catalogs, it was probably made by another firm for Arcade, possibly during the "toy trust" of the early 1900s, as it is unlike any other of Arcade's limited line of ranges. The vertical backpiece resembles the small "Tot" by Grey Iron, and the cast iron panels resemble Shimer's similar offerings. Arcade catalogs also show this "Gem" range without a left hearth shelf. 3" l., mfgr. Arcade or Shimer. $50-75.

147. Gem: Small range without cast-in hearth shelf. Although Arcade Mfg. Co. sold this small "Gem" range in their catalogs, it was probably made by Wm. Shimer & Son. The design on all cast sides is identical to Shimer's "Nit'," as well as the "Cute," "Pet," "Dot," and "Home" of the same size and series. 3" l., mfgr. Shimer or Arcade. $50-75. *Courtesy of Tiny Moyer.*

148. Geneva: Small gas range, catalog no. 545. Champion Hardware Co. only produced this small toy range, and a larger version thereof (no. 445), in red, green, and blue, plus the small upright style "Daisy" range. 3 1/4" l., mfgr. Champion, Geneva, Ohio. $75-100.

149. Globe: The third largest range produced by The Kenton Hardware Co. No. 950. 22 7/8" l., mfgr. Kenton. $1,800-2,200. *Courtesy of Tiny Moyer.*

150. Godin: European style cast iron range: "BTE S.G.D.G." 13" l., mfgr. Godin & Cie. $600-750. *Photo courtesy of Ralph and Merlyn Wiesehan.*

151. Godin: Porcelainized French range. 11" l., mfgr. Godin & Cie. $750-850. *Photo courtesy of Ralph and Merlyn Wiesehan.*

152. Gold Coin: Nickel finish range. 15 1/4" l., mfgr. Bussey & McLeod Stove Co. $1,500-1,750. *Courtesy of Tiny Moyer.*

153. Grand Jewel: Ornate nickel finish range. This range was also made with a vignette of a lady on the oven door. The extension shelf design is the same as the small size. 19" l., mfgr. probably Ideal. $1,500-2,000. *Courtesy of Tiny Moyer.*

154. Grand Jewel: c. 1903. Nickel range with backpiece offset to the left side and stovepipe to the right. The extension shelf sets on two pins, which is unusual. This offset backpiece feature is typical of Ideal ranges. Note the similar "Pet" and "Perfection" of this size range. The missing oven door is the same as Ideal used on the "Pet" range of this size. 10 1/2" l., mfgr. Ideal. $250-300.

155. Great Majestic Junior (The): Large nickel and black finish range, with side reservoir. 26" l., mfgr. Majestic Manufacturing Co., St. Louis. $4,000-5,000. *Courtesy of Tiny Moyer.*

156. Great Majestic Junior (The): Large nickel and black finish range, with side feeder. 23 1/2' l., mfgr. Majestic Manufacturing Co., St. Louis. $4,000-5,000. *Courtesy of Tiny Moyer.*

157. Great Majestic Junior (The): Large nickel and black finish range. 24 1/2" l., mfgr. Majestic Manufacturing Co., St. Louis. $4,000-5,000.

H

158. Hanks Junior: Black finish, with three nickel plated doors and two loose trivets suspended from low backpiece. This range is similar in design to the "Wetter," "U. S." and "Mascot" produced by the U. S. Stove Corp. of South Pittsburg, Tenn., formerly H. Wetter Mfg. Co./Bluff City Stove Works of Memphis, Tenn., and the Mascot Stove Mfg. Co. of Chattanooga, Tenn. The "Baby Hazel" is a similar range, marked "Rome Stove & Range Co., Rome, Georgia." 17 1/2" l., mfgr. Hanks Stove and Range Co., Rome, Georgia. $3000-3,500.

159. Holly: Nickel finish range with nicely engraved hearth shelf. 7 1/2" l., mfgr. J&E Stevens . $250-350. *Courtesy of Bruce and Melinda Kling.*

160 (not shown). Home: Nickel plated range with "lady with bonnet" design in front base skirt. The hearth shelf is marked "Home," and the backpiece is ornate. This range has been reported, but not verified. The "lady with bonnet" feature is like the base of the "Princess" and "Superior" ranges listed herein. The ornate backpiece is like one reported on a "Champion" range of similar size. 15 1/2" l., mfgr. Ideal. $500-650.

161. Home: Largest size of the Stevens line of "Home" ranges, with nickel finish and simulated clock design above name on oven door. No. 67. None of Stevens' "Home" ranges appear in their 1906 catalog, but nine sizes and types of "Home" ranges are shown in their 1924 catalog, varying in length from 14 1/2" down to 7 1/2". 14 1/2" l., mfgr. J&E Stevens. $500-750.

163. Home: Nickel finish range, with simulated clock in center of oven door. 11 1/2" l., mfgr. J&E Stevens. $350-450.

162. Home: Electro oxidized finish range, with center simulated clock design on oven door. No. 166. None of Stevens' "Home" ranges appear in their 1906 catalog, but nine sizes and types of "Home" ranges are shown in their 1924 catalog, varying in length from 14 1/2" down to 7 1/2". 13" l., mfgr. J&E Stevens. $400-600.

164. Home: Nickel finish range No. 62. The 1924 Stevens catalog lists this, the smallest, of their "Home" ranges, as 7 1/2" length with a hearth shelf. However, this actual example measures 8 1/4". 8 1/4" l., mfgr. J&E Stevens. $250-350.

166. Home: Electro oxidized range with name on oven door within a heart shape, and also on hearth shelf. No. 4. Similar size small ranges, with the heart shape on the oven door, are also shown in Kenton's catalog. The same "Queen" version is shown in both Kenton's and Grey Iron's catalogs. Interestingly, the Grey Iron listing shows a Kenton lid lifter. As a result of the "toy trust" in the early 1900s, the various toy firm members made parts and complete toys for each other. Grey Iron also produced a cheaper version of this range with a tin left side panel and no hearth shelf. 5 7/8" l., mfgr. Grey Iron. $125-150.

165. Home: Nickel finish range, shown as No. 915 in Kenton's 1905 catalog. This item was produced into the 1920s. 7" l., mfgr. Kenton. $125-150.

167. Home: Nickel finish range, No. 3 1/2, with fluted borders around the base and stove top. This "Home" range was also sold with just a stovepipe, instead of a removable backpiece, and plain borders around the base and stove top, as catalog No. 3. 5" l., mfgr. Grey Iron. $125-150.

168. Home: Small nickel range with all cast sides and removable backpiece. This is one of a family of such small ranges with integrally cast hearth shelf and underneath turnkey take-apart feature. Names on the oven door are "Cute," "Pet," "Dot," "Gem" (sold by Arcade), and Shimer's "Nit'" (less the hearth shelf). 3 1/4" l., mfgr. Shimer. $85-125.

169 (not shown). Hot Plate: Electric hot plate, green, two burners. This electric hot plate could be used for cooking purposes, but was included in Arcade's laundry set, No. 684, in the 1930s. 2 3/4" l., mfgr. Arcade. $50-75.

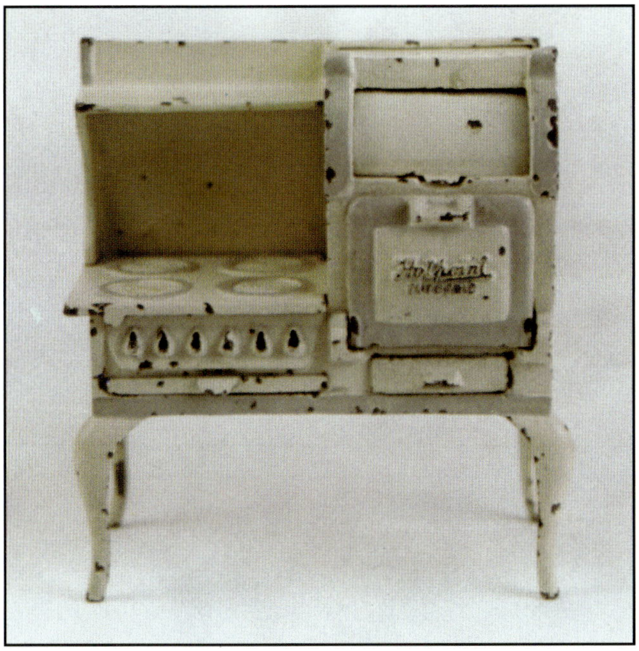

170. Hotpoint: Electric range painted white with gray trim. This range was also made as a bank version. 5 1/4" l., mfgr. Arcade. $200-300. *Courtesy of Tiny Moyer.*

171. Hotpoint: Copper finish electric range. This is the plated "specimen" pictured in Al Aune's book entitled "*Arcade Toys,*" page 124, and was sold in the auction of Al's Arcade toy collection after he passed. Production examples of this electric range are painted. 5 1/2" l., mfgr. Arcade. $300-400.

172. (not shown) Hotpoint: Smaller size Hotpoint gas range; this is a case where the smaller size of a range is considerably rarer than the larger size. 3" l., mfgr. Arcade. $200-300.

173 (not shown). Hotpoint "The Symphony" Range Bank: White range bank trimmed in black. A rare still bank, which has been noted without the wording "The Symphony." 3 7/8" l., mfgr. Arcade. $600-800.

174. Hotpoint: Green, two-burner heater stove with accompanying copper plated boiler. Although this little heater stove was sold in Arcade's laundry room set, c. 1929, it could also be used for warming or cooking purposes. 2 5/8" l., mfgr. Arcade. $50-75.

175. Husqvarna No. 2: Large black finish Swedish toy range with pot and pans. 14 1/4" l., mfgr. Husqvarna (Sweden). $1,000-1,500.

☙ I ❧

175A. Ida: Range with black finish and a scalloped base. This range is identical to the "Dainty" range produced by Orr, Painter & Co. except for the names thereon. It has a center stovepipe and extension shelf, but no heath shelf. 12 1/2 " l., mfgr. Southern Stove Works Co., Richmond, VA. $1,250-1,750.

177. Ideal No. 3: Large nickel finish range, c. 1894. 18" l., mfgr. Ideal. $1,200-1,400. *Courtesy of Bruce and Melinda Kling.*

176. Ideal No 5: Large nickel finish range, c. 1894. 23 1/2" l., mfgr. Ideal. $1,500-1,750.

178. Ideal No. 2: Copper flash finish with backpiece offset to the left and separate stovepipe to the right. c. 1894. 11" l., mfgr. Ideal. $750-1,000.

179. Ideal No. 1: Nickel finish range, c. 1894. 7 3/4" l., mfgr. Ideal. $350-450.

180. Ideal Still Bank: Small nickel finish gas range still bank. This bank has the wording, "The Ideal Mfg. Co. Detroit, Mich." and "Compliments of the Ideal Mfg. Co." 3 1/8" l., mfgr. Ideal. $1,500-1,750.

181. Indiana: Cook stove. 10 1/2" l., mfgr. Charles Kibler Jr.& Co. Newark, Ohio. $750-1,000. *Photo courtesy of Ralph and Merlyn Wiesehan.*

182. Iron Is Livitamin: Modern modification of the original "Queen" of this design, for advertising. The original "Queen" in nickel has been voluminously reproduced with a black finish, and the door has been further modified to reflect advertising. See next listing for comparison to original "Queen" range. 5" l., mfgr. Grey Iron Modification. $25-50. *Courtesy of Tiny Moyer.*

183. Iron is Livitamin: Black finish modification on original nickel finish "Queen" range for advertising . This advertising item is shown in the same photo with an original "Queen" range. 5" l., mfgr. Grey Iron. $25-50, & $150-200. *Courtesy of Tiny Moyer.*

184. Ivy: Nickel finish range, made with hearth shelf only. Cat. no. 51. 6 3/4" l., mfgr. J&E Stevens. $250-300.

185. I X L: Small nickel plated range, produced with left side hearth shelf, but no extension shelf. An "I X L" iron serves as a massive "go-with" for this toy range. 8" l., mfgr. Ideal. $125-175.

ℭ J ℘

186. James Leffel's: Cook stove marked "Double Oven Stove. Pat. 1849 Springfield, O." Size not available, mfgr. James Leffel. $750-1,000. *Photo courtesy of Ralph and Merlyn Wiesehan.*

187. Jean: Brass pattern for the "Jean" gas range. This pattern for the "Jean" range was located with a group of other Dent Hardware Co. patterns. Although this range is, in fact, listed in Dent's "Catalogue No. 10" on page 38, as well as a smaller 3 7/8" size, no actual production specimens have thus far been seen by the authors. 4 7/8" l., mfgr. Dent. $500-750.

189. Jewel Range Jr.: Ornate nickel plated range, with water reservoir. 18 1/2" l., mfgr. Detroit Stove Works. $1,500-1,800.

188. Jersey: A beautiful, large, nickel and black finish range. Although no mfgr.'s name is stated, there are notable similarities between this range and the "St. Clair" listed herein. 16" l., mfgr. uncertain. $4,500-6,000. *Courtesy of Tiny Moyer.*

190. Jewel Range Jr.: Ornate nickel plated range, made without the side water reservoir. 14 7/8" l., mfgr. Detroit Stove Works. $800-1,000. *Courtesy of Tiny Moyer.*

191. Jewel Range Jr.: Ornate nickel plated range, with company crest on the oven door, and marked "Detroit Stove Works" and "The Original Detroit Stoves." This later style "Jewel Range Jr." was also produced with a reservoir on the right side. 15 1/4" l., mfgr. Detroit Stove Works. $1,500-1,850. *Courtesy of Tiny Moyer.*

193. Julia: Four-burner stove, with black finish. 11 5/8" l., mfgr. Philadelphia Stove Works. $500-750.

192. Jewel: Two-burner small range with correct "stew-pan, skillet, and lifter," as sold. Catalog No. 5 1/2 X is nickel plated and No. 5 1/2 XC is "finished in colored lacquers," sold "3 colors in a box." This range takes the second to the smallest size pot and pan, which are difficult to find and are shown on the yellow version in the photo. *Note:* the black finish version may be a later reproduction. 4 3/4" l., mfgr. Grey Iron. $125-150.

194. Jupiter: Largest of Shimer's ranges, with ornate castings. Two distinct designs of the "Jupiter" were produced by Shimer. 14" l., mfgr. Shimer. $800-1,200.

195. Jupiter: Largest of Shimer's ranges, with ornate castings with star and crescent motif. Two distinctive designs of the "Jupiter" were produced. 14 3/4" l., mfgr. Shimer. $1,200-1,400. *Courtesy of Tiny Moyer.*

☙ K ❧

196. Karr: Original range with blue porcelain and nickel finish. This is the original version. Reproductions of this original were produced in various colors in the 1960-1970 period and also later. 12 3/4" l., mfgr. Karr Range Co., Belleville, IL. $3,000-4,000.

197. Karr: Reproduction in red of the original Karr range of this design. Reproductions of the original were produced in various colors, c. 1960s, and also later. 12 1/4 " l., mfgr. uncertain. $1,000-1,500. *Courtesy of Tiny Moyer.*

198. Kent: Largest of Kenton's new line of gas ranges for 1927. Produced in royal blue and nickel. The 10" and 8" sizes were also available without the third, upper oven, as well as in 6 1/2" and 5" sizes. 10" l., mfgr. Kenton. $500-600.

199. Kenton "Table Top Stove." New for 1936, this green and nickel gas range has a removable cover on the left top covering all four burners. No. 4040. (Beware of modern reproductions.) 4" l., mfgr. Kenton. $300-400.

200 (not shown). King: Small nickel plated range, No. 43. 4" l., mfgr. Shimer. $150-200.

L

201. Leader No. 649: Large, well detailed range, with porcelain and nickel finish. This item may well be a "salesman's sample" rather than a toy range due to its quality and detail. 20 1/4", mfgr. Washington. $4,000-5,500. *Courtesy of Tiny Moyer.*

203. Libertas: Copper finish range with scrollwork design on the oven door. 13 1/2" l., mfgr. uncertain, but probably Ideal. $400-500.

202 (not shown). Leader: No further details available at time of printing. However, this is a mid-size range with a stovepipe, similar to the "Libertas" and "Liberty" listed herein. Mfgr. uncertain, but probably Ideal. $400-500.

204. Libertas: Nickel and black finish range, with bust of an eagle on the oven door. This same basic range has been observed with a copper finish and the name "Liberty" on hearth plate. 13 1/2" l., mfgr. uncertain, but probably Ideal. $400-500. *Courtesy of Tiny Moyer.*

205. Liberty: Nickel finish range, similar to the "Libertas" range. This range has also been noted with a copper flash finish, and an eagle bust and name "Libertas" on the oven door. 13 1/2" l., mfgr. uncertain, but probably Ideal. $400-500.

206. Lilly: Four-burner stove with black finish. The production patterns for this stove must have passed from the Philadelphia Stove Works to Orr, Painter & Co., as the "Julia" stove listed herein is identical, even the lid lifter. 12" l., mfgr. Orr Painter & Co. Reading, PA. $500-750.

207. Little Cod: Black finish range, with cod fish decoration. This range is included herein due to its similar size to a toy range, but it is believed to have been actually used on fishing boats. 17 1/2" l., mfgr. Charles Fawcett Mfg. Co. Ltd., Sackville, N.B., Canada. $1,200-1,500. *Courtesy of Tiny Moyer.*

209. Little Ethel: Black finish stove with same basic design as the "Little Willie" stove shown herein, but has thinner base and shorter legs. 13 7/8" l., mfgr. uncertain, perhaps Abbott & Noble, or its successor. $500-750. *Courtesy of Tiny Moyer.*

208 (not shown). Little Daisy: Range with black finish and nickel plated oven door. Since the design on this range is similar to the "Royal Esther" range, the manufacturer was possibly Mt. Penn Stove Works of Reading, PA. Approximately 16" l., mfgr. uncertain. $2,500-3,000.

210. Little Eva: Early black finish stove, with maker's name on hearth plate, removable legs. The Boston Almanac records N. S. Cate at 25 Sea Street in 1847, but by 1848, Earl N. Cate is recorded at this address. This stove is also known with the makers' names of J.W. Clark, Scranton Stove Co., T. Southard, and G.F. Filley, which indicates that toy stove patterns were transferred from one mfgr. to another. This stove design is known to have either three or four legs, and are either attached or removable. 14" l., mfgr. N.S. Cate, Boston. $800-1,000 (this type). $500-700 (later type).

212. Little Giant: Black finish stove, with four burners, and marked "G.D. Harris, Pattern Maker, CIN." 11" l., mfgr. Harris, G.D. $800-1,000. *Photo Courtesy of Tiny Moyer.*

211. Little Fanny: Black finish stove. Hearth plate reads "Little Ida Smith." 11 3/4" l., mfgr. Philadelphia Stove Works. $500-750.

213. Little Willie: Early black finish stove, with thick base and high legs. A later version of this stove, also by Abbott & Noble, has their name cast on the front hearth plate instead of along the side, with a much thinner base, and shorter legs. 14" l., mfgr. Abbott & Noble, Philadelphia. $800-1,000.

214. Loth's Fuel Saver: Small blue and ivory painted range, with upper warming oven, and advertising Loth. The W.J. Loth Stove Co. was located in Waynesboro, Virginia. 4 1/4" l., mfgr. Arcade. $450-600. *Courtesy of Tiny Moyer.*

M

215. Magic Chef: Gas range die cast bank, painted white with black trim. This advertising range is shown with its original box and Magic Chef standing figure. Although this gas range is made of pot metal, rather than cast iron, it does fit in nicely with related cast iron ranges. 2 7/8" l., mfgr. National Products Corp., Chicago. $75-150.

217. Marvel: Nickel plated version of Kenton's fourth largest range. This range was also produced with a different style backpiece containing two warming ovens. 19 1/2" l., mfgr. Kenton. $450-650.

216 (not shown). Mars: Small range, with left side hearth shelf and stovepipe at the right rear. This range was produced with either a black, or nickel plated finish. Cat. no. 150. 5 1/4" l., mfgr. Shimer. $150-200.

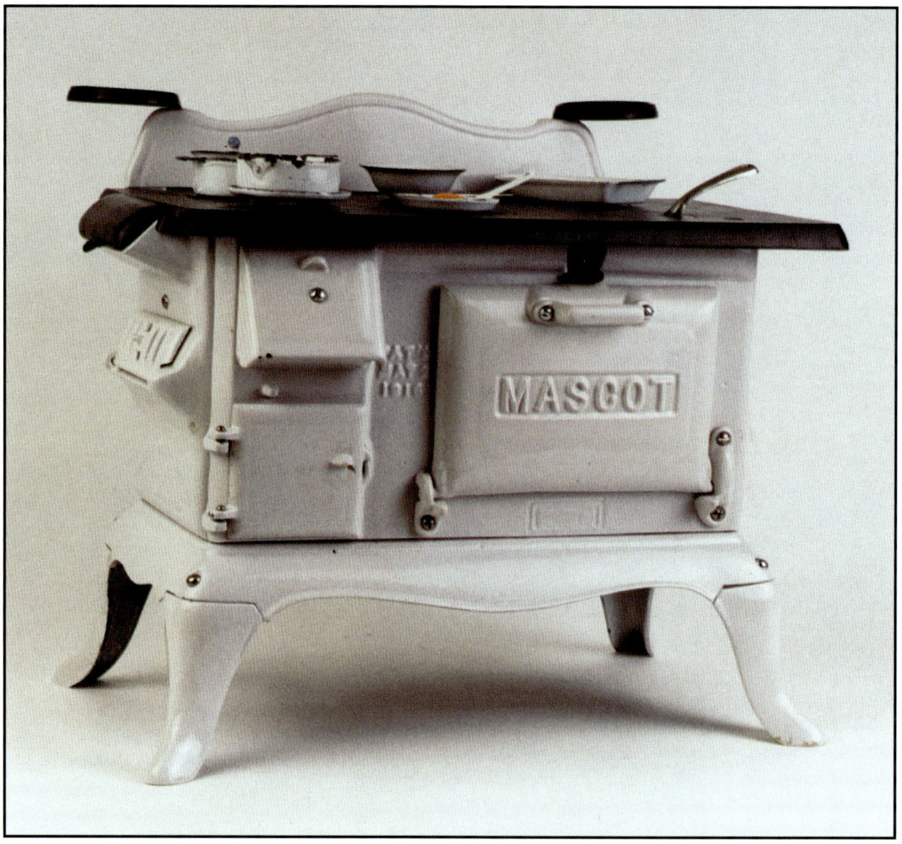

218. Mascot: Range with white porcelainized finish. Patented. May 2, 1916. This range is similar in design to the "Wetter," "U.S." and "Hanks Junior," which is indicative of the progression of stove/range manufacturers utilizing the same basic patterns over a span of many years. 17 7/8" l., mfgr. Mascot Stove Mfg. Co., Chattanooga, Tenn. $3,000-3,500. *Courtesy of Tiny Moyer.*

220. Mechanic: Early three step stove, with black finish and nickel plated oven door. Pat'd. 1860. 12" l., mfgr. Brecher, P.&G. $800-1,000. *Photo Courtesy of Tiny Moyer.*

219 (not shown). May: No further details at time of printing.

221. Mercury: The second largest size of Shimer's range offerings, in nickel finish. 10 1/2" l., mfgr. Shimer. $500-600.

222 (not shown). Merry Christmas Ida: Early black finish stove, with maker's name on hearth plate. Generally the same toy stove as the "Little Eva", by N. S. Cate in Boston, except the foundry patterns must have subsequently been transferred to the Scranton Stove Co. The girl's name "Ida" must have had some special significance, such as the daughter of the firm's owner. Note that the name "Little Ida Smith" is on the removable hearth plate on the "Little Fanny" stove by the Philadelphia Stove Works. Furthermore, the name "Ida" is on the oven door on the range listed as No. 175A herein. 14" l., mfgr. Scranton Stove Co. $500-750.

223. Merry Christmas: Black finish stove with four legs. Similar to the three-legged "My Favorite" stove. 11 5/8" l., mfgr. uncertain, (possibly T. Southard or G.F. Filley) $500-750. *Courtesy of Tiny Moyer.*

225. Midget: Nickel plated range. Similar design to Ideal's "Baby" range. 14 1/2" l., mfgr. Ideal. $600-800. *Courtesy of Tiny Moyer.*

224 (not shown). Midget: Large upright range, with full-length upper warming closet. This range is similar to the "St. Clair" range shown herein. 16" l., mfgr. Belleville Stove Works, Belleville, IL. $5000-7,500.

226. Mills Range Jr.: Large, elaborate nickel plated range. This range is dated 1881, and has a separate base. When the oven door is opened, a movable scrollwork shelf lowers, and again raises upward when the door is closed. M.I. Mills was a major financial backer of The Michigan Stove Co. 20" l., mfgr. The Michigan Stove Co. $2,500-3,500.

227. Model Enterprise No. 40: Large black stove, with movable shelves and two trivets. Two "go-withs" issued for advertising by this mfgr. (pie pan and skillet) are shown. 22 1/2" l., mfgr. Phillips & Buttorff Mfg. Co., Nashville, Tenn. (est. 1881). $2,500-3,500.

228. Model: Kenton's largest toy range, with electro oxidized finish, and complete with all the correct accessories. This range was also produced with a nickel plated finish. 26 1/2" l., mfgr. Kenton. $2,500-3,500.

229. Monarch: Deluxe quality porcelainized range. The Monarch miniature range may have also been utilized for "salesman's sample" purposes. 24" l., mfgr. Malleable Iron Range Co. $4,500-5,500. *Photo courtesy of Donna Johnson.*

230. My Favorite: Black finish stove with three legs. Similar to the four-legged "Merry Christmas" stove. 11 5/8 " l., mfgr. uncertain, possibly T. Southard or G.F. Filley. $500-750. *Courtesy of Tiny Moyer.*

231. My Joy: Range with blue enamel and nickel finish. The catalog No. is 60 1/2 EB. This range was also produced nickel plated (No. 60 1/2), without top shelf (No. 60 1/4) and with no shelves (No. 60). 11 1/4" l., mfgr. Grey Iron. $600-750.

૯ N ૱

232 (not shown). National Journal: Modern black finish range. It appears this range was recently produced, using mainly patterns once used by Grey Iron Casting Company, with a modified oven door stating the name. Approx. 10" l., mfgr. uncertain. $10-15.

233. National: This "National Steel Range" has the removable extension shelf on the left side and water reservoir on the right, with a full upper warming oven, or "closet". Cat. no. 995. This attractive cast iron and black finish steel range was advertised in the Kenton Hardware Mfg. Co. catalog of 1905 for $2.00. It is similar in design, but smaller than Kenton's "Novelty" range. A cheaper version of the "National" was offered without the upper warming oven, as cat. no. 900. The example illustrated has been restored. 13" l., mfgr. Kenton. $600-800.

234. New Empire: Cookstove dated 1872. 14" l., mfgr. Sweet, Quimby & Perry, Troy, N. Y. $1,200-1,500. *Photo courtesy of Ralph and Merlyn Wiesehan.*

235. New England Kitchen: Range with black finish and gold highlighting, catalog No. 200. This range is shown in Wilkins' 1895 catalog, but not in their 1911 catalog. 14 1/2" l., mfgr. Wilkins. $1,200-1,500. *Courtesy of Tiny Moyer.*

237. Nit': Small nickel plated range, marked "40." An underneath turnkey dismantles this small range. All sides are cast iron, with ornamental design. 3 1/4" l., mfgr. Shimer. $85-125.

236. Nineteen Hundred (1900): Small, two-burner nickel plated range with stovepipe. The "1900" is very similar to Kenton's "Zoo" range. 4 1/4" l., mfgr. Kenton. $100-125. *Courtesy of Bruce and Melinda Kling.*

238. N.O: European style range, probably French. 17 1/2" l., mfgr. uncertain. $600-750. *Courtesy of Tiny Moyer.*

239. Norco: Large nickel plated range, with two side shelves and backpiece surrounding the stovepipe. The design of the "Norco" so closely resembles Hubley's "Eagle" range of this similar length, that the strong probability is that Hubley produced this range as a private label arrangement. 19 1/2" l., mfgr. probably Hubley. $500-750. *Photo Courtesy of Tiny Moyer.*

240. Novelty: Deluxe version of this black and nickel plated range which features a full upper warming oven and extension shelves on both sides. Catalog no. 985. Two cheaper versions of the "Novelty" range were offered. No. 980 had no extension shelves nor upper backpiece warming oven , and No. 975 had no backpiece at all and no shelves, just a stand alone stovepipe. Beware of modern reproductions. 12 1/2" l., mfgr. Kenton. $450-750.

O

241. Oak: Range produced in the 1890s. Front and left side cast panels are very similar to Kenton's "Eva" range. 8 1/2" l., mfgr. Kenton. $300-400.

243. O K: Small range shown in two versions, both with stovepipe. Both a nickel plated version and a copper finish version are shown in the same photo, plus an advertising cast iron hinged facsimile of a range with "OK Stove and Range Co." under the "cooking surface." The version without a shelf is offered in the 1892 Marshall Field & Co. catalog. 6 1/4" l., & 5" l., mfgr. Ideal. $125-175.

242 (not shown). Oak: Small nickel plated range, No. 38. Made as a still bank, also. 2 3/8" l., mfgr. Shimer. $100-150.

244. Our Baby: Canadian made six-burner range, with mfgr.'s name embossed on the left side hearth plate. 15" l., mfgr. Standard Foundry Co., Montreal, P. Q., Canada. $600-750. *Photo Courtesy of Tiny Moyer.*

245. Our Pet: Black finish stove, with nickel door. An advertising "go-with" is also shown in this photo for Bridgeford & Co. 11 1/2" l., mfgr. Bridgeford & Co., Louisville, KY. $1,500-1,750.

P

246. Passion Flower: Cook stove marked "1886, Evansville, Ind." Made both nickel plated and porcelainized. 20" l., mfgr. Southern Stove Works, Evansville. $2,500-3,500. *Courtesy of Tiny Moyer.*

247. Passion Flower: Cook stove marked "1886, Evansville, Ind." Made both nickel plated and porcelainized. 20" l., mfgr. Southern Stove Works, Evansville. $2,500-3,500. *Photo courtesy of Ralph and Merlyn Wiesehan.*

248. Pearl: Small range with copper finish and the name within a heart on the oven door. However, the side hearth shelf reads "Home." Similar small ranges of this size, with the names "Queen" and "Home" within the heart shaped design on the oven door are shown in a Grey Iron Casting Co. catalog. It appears Kenton produced this line of "heart design" small ranges, and some were also marketed by Grey Iron. 5 5/8" l., mfgr. Kenton. $125-175.

249. Pearl: Name within "heart" on the oven door. Both four-burner and two-burner varieties are shown and both are nickel plated. 5" l., & 4 3/4" l., mfgr. Kenton. $125-150.

251. Pearl: Small nickel plated range with stovepipe only. 5" l., mfgr. Kenton. $100-125.

250. Pearl: Small nickel plated range, with two shelves and removable back piece. 7" l., mfgr. Kenton. $150-200.

252. Peninsular: White painted range with gray trim, shown with several advertising "go-withs" for the Peninsular Stove Co. Initially sold in 1927. 7 1/4" l., mfgr. Arcade. $600-750.

253. Perfection: Range with nickel and black finish and six pointed star on oven door. Various sizes of "Perfection" ranges were later produced by Kenton. There are two additional varieties of this particular "Perfection" range. One has an ornate center backpiece with a five-pointed star at the top. The other has an even more elaborate backpiece, which includes the name "Perfection", and also an extension shelf on the right side, which extends the total length to 18". 14" l., mfgr. Ideal or Kenton. $350-450.

254. Perfection: Nickel plated range, with backpiece offset to left. The backpiece offset to the left is typical of many ranges by "Ideal," such as the "Ideal No. 2," "Pet," and "Grand Jewel." Similar "Perfection" ranges were later produced by Kenton. 7 1/2" l., mfgr. Ideal. $250-300.

255. Perfection: Nickel plated range, missing the stovepipe. This range was also offered with a backpiece in 10 1/4" & 8" lengths. 10 1/2" l., (also offered in 14", and 7 3/4" lengths) mfgr. Kenton. $250-350. *Courtesy of Tiny Moyer.*

256. Perfection: Small four-burner range, made without side shelves. Copper flashed finish. Kenton often produced both copper flash and nickel plated finishes on the same range. 4 3/4" l., mfgr. Kenton. $100-125.

258. Pet: Copper flash range, with hearth shelf and stovepipe. This same range has been seen marked both with "Ideal" or "Kenton" on the back. Also, the burner lids are different. 11" l., mfgr. Ideal and Kenton. $300-400. *Courtesy of Tiny Moyer.*

257. Pet: Black finish stove, with removable hearth plate and no other markings. 14 1/2" l., mfgr. uncertain. $500-750. *Courtesy of Tiny Moyer.*

259. Pet: Nickel plated range, with copper finish oven door. The offset backpiece is to the left, and stovepipe is to the right. The backpiece offset to the left is typical of many ranges by "Ideal," such as the "Ideal No. 2," and "Grand Jewel." 7 1/2" l., mfgr. Ideal. $250-300.

262. Pride: Nickel plated range, missing the left side hearth shelf. Similar in design to Kenton's "Royal" range shown in this size, and Hubley's "Eagle" range of this size, which is illustrative of the toy industry's practice of making pattern tooling, parts, and even complete items for each other. The base is similar to Dent Hardware Co.'s work. 5 1/2" l., (as is), mfgr. uncertain, possibly Dent or Kenton. $250-350.

260. Pet: Small nickel plated range with all cast sides and removable backpiece. This is one of a family of such small ranges with integrally cast hearth shelf and underneath turnkey take-apart feature. Names on the oven door are "Cute," "Home," "Dot," "Gem" (sold by Arcade), and Shimer's "Nit'" (made less the hearth shelf). 3 1/4" l., mfgr. Shimer. $85-125.

261. Plymouth Rock: Cookstove. 9 1/2" l., mfgr. G.F. Filley, St. Louis. $800-1,000. *Photo courtesy of Ralph & Merlyn Wiesehan.*

263. Princess: Electro oxydized range with two removable shelves and backpiece. "lady with bonnet" design both on oven door and on center of front base skirt. The "lady with bonnet" design is common to the Ideal "Superior" ranges listed herein. A further design variation has three horses heads on the oven door design, and the "lady with bonnet" in the base skirt. However, the smaller Ideal "Princess" range does not contain the "lady with bonnet" design. The extension shelf (tipped upward for easier viewing in the photo) is common to several Ideal ranges. 15 1/2" l., mfgr. Ideal. $600-800.

263A. Princess: Electro oxidized range, which requires two side shelves and a backpiece. Three horses heads are on the oven door, and the "lady with bonnet" design, which is common to other Ideal ranges, is in the front base skirt. 15" l., (if complete) mfgr. Ideal $600-800. *Photo Courtesy of Tiny Moyer.*

264. Princess: Range with two shelves and painted aluminum. This range has a typical "Ideal" design with ornamental pattern on the oven door and the backpiece offset to the left, and the stovepipe to the right. Two other similar ranges of this design are named "Queen" and "Superior" on the left side removable hearth shelf. 12 1/2" l., mfgr. Ideal. $400-500.

265. Princess: Nickel plated range with typical "Ideal" design pattern on the oven door, and the placement of the backpiece offset to the left, and stovepipe to the right. 10 3/4" l., mfgr. Ideal. $300-400.

267. Prize: Nickel plated, ornate range. 13 1/2" l., mfgr. J&E Stevens. $400-500.

266 (not shown). Princess: Modern issue range with black finish, and gold lettering on the oven door. Same design as Grey Iron Casting Co.'s "Queen" range reproduction shown herein, but smaller scale. 3" l., mfgr. Taiwan. $10-15.

Q

268 (not shown). Qualified: Nickel plated and tin range, with full length upper warming oven. Other than name, this range appears identical to the "Karr" range shown herein. 12 3/4" l., mfgr. Qualified Range Co., Ft. Recovery, Ohio. $2,500-3,000.

269. Queen (The): Brass pattern for Dent's largest size "The Queen" range, with elaborate backpiece containing two upper warming ovens. 24 1/2" l., mfgr. Dent. $2,500-3,000.

270. Queen (The): Dent's largest size "The Queen" range, with copper flash finish. Correct utensils and lid lifter are shown with this range. Two backpiece options were offered for this range, the ornamental version shown, and a more elaborate version with two upper warming ovens shown in the pattern photo. 24" l., mfgr. Dent. $4,000-5,000.

271. Queen (The): Dent's largest size "The Queen" range with dual finish of nickel and copper flash. No right hand extension shelf was included on this model. 20 1/2" l., mfgr. Dent. $4,000-5,000. *Courtesy of Tiny Moyer.*

272. Queen (The): Dent's second to largest size "The Queen" range with copper flash finish. 18 3/4" l., mfgr. Dent. $1,500-2,000. *Courtesy of Tiny Moyer.*

273. Queen (The): Dent's second to largest size "The Queen" range, with nickel plated finish. 18 3/4" l., mfgr. Dent. $1,500-2,000. *Courtesy of Tiny Moyer.*

274. Queen (The): Dent's third largest size "The Queen" range, with nickel plated finish. No right-hand extension shelf was included on this model. 14 3/4" l., mfgr. Dent. $1,200-1,500. *Courtesy of Tiny Moyer*

276. Queen (The): Mid-size version of "The Queen," with copper flash finish. 11" l., mfgr. Dent. $250-350. *Courtesy of Tiny Moyer.*

275. Queen (The): Brass pattern for Dent "The Queen" range. 15" l., mfgr. Dent. $1,000-1,200.

277. Queen (The): Mid-size version of "The Queen" with nickel plated finish. 11" l., mfgr. Dent. $250-300. *Courtesy of Tiny Moyer.*

278. Queen: Brass pattern for Dent "Queen" range. 10 1/2" l., mfgr. Dent. $500-750.

279. Queen: Brass pattern for Dent "Queen" range. 8 1/2" l., mfgr. Dent. $400-600.

280. Queen: Range with nickel plated finish and stovepipe. Shown with bottom section of the original box. 8" l., mfgr. Dent. $150-200.

281. Queen: Range with copper flash finish, and ornate removable backpiece. 7 1/4" l., mfgr. Dent. $150-200.

282. Queen: Both a four-burner, with hearth shelf, and a two-burner example of the small Dent "Queen" range are shown in this photo. 6 1/4" & 4 3/4" l., mfgr. Dent. $100-125. & $75.-100.

283. Queen: Brass pattern for the smallest size Dent "Queen" range, shown with correct pot and pan. 4 3/4" l., mfgr. Dent. $250-300.

284. Queen: Mid-size, nickel plated range, with ornamental overall design typical of Stevens' quality offerings. 13" l., mfgr. J&E Stevens. $400-500.

285. Queen: Black finish range with nickel trim. Typical "Ideal" design, with ornamental door pattern, and backpiece offset to the left. There are three ranges utilizing the same basic design, which are identified as "Queen," "Princess," and "Superior" on the left side removable hearth shelf. 12 1/2" l., mfgr. Ideal. $400-500.

286. Queen: Range with blue enamel and nickel finish. Catalog No. 12 1/2 EB. This small range was sold with and without utensils, also nickel plated or with a black finish. Reproductions, in black, by superseding firms are extremely prolific and commonly seen in "antique" malls and on the Internet. 6 1/4" l., mfgr. Grey Iron. $200-300.

287. Queen: Range with nickel finish and original box. However, the box is marked "No. 20," which is the smaller size "Crescent" nickel plated range made by this firm. This range is catalog No. 12 1/2 N. The nickel version is slightly shorter in length than the enameled version of this range. Reproductions, in black, by superseding firms, are extremely prolific and commonly seen in "antique" malls and on the Internet. 6" l., mfgr. Grey Iron. $175-250.

288. Queen: Reproduction made from original patterns by Grey Iron, with original box, and sold under the "Greycraft" name. This officially declared reproduction by Grey Iron should not be confused with the extremely common reproductions subsequently made by other firms. The box for a later repro is marked "Albert E. Price, Inc., Item No. 9296, Taiwan," c. 1970s. 6" l., mfgr. Grey Iron. $50-75.

290. Queen: Modern die cast range, with right side entry for pencil sharpening. Copper finish. Similar, miniature design characteristics to Grey Iron Casting Co.'s "Queen" range and subsequent reproductions thereof. 2 1/2" l., mfgr. uncertain. $10-15.

289 (not shown). Queen No. 4: Word "Queen" within heart shape on oven door, and also on removable left side hearth shelf. This four-burner model is shown in Grey Iron's catalog with Kenton style lid lifter, plus stovepipe and no backpiece. This heart shaped design on the oven door, with range name within, is the same basic design as the "Home" and "Pearl" ranges of this size shown in both Kenton and Grey Iron catalogs as "No. 4." 6" l., (per catalog), mfgr. Grey Iron & Kenton. $125-150.

292. Quick Meal: Large gas range. 14 1/2" l., mfgr. Quick Meal Stove Co. Div., American Stove Co. $1,500-1,750.

291. Quick Meal: Restored example of the "Quick Meal 407-16." The correct lid lifter is also marked "Quick Meal." 17 1/2" l., mfgr. Quick Meal Stove Co. Div., American Stove Co. $2,800-4,000. *Courtesy of Tiny Moyer.*

293. Quick Meal: Small painted and nickel plated version of this range, which is the only size offering by Arcade. 5 1/8" l., mfgr. Arcade. $750-1,250.

R

294 (not shown). Ranch Craft: Modern made range, with black finish and two side shelves, plus backpiece with stovepipe. The logo on the oven door is a circle containing the letters "RC." This modern range appears to be made for gift shops, or home decoration. Approx. 9" l., mfgr. uncertain. $15-25.

295. Range Eternal (The): Large toy range modeled closely after this firm's full-size range. 21 1/2" l., mfgr. Engman-Matthews Range Co. $3,800-5,500.

297. Richmond Model: Large highly detailed range. Due to its exacting detail, this range could originally have served as a salesman's sample. 18 7/8" l., mfgr. Richmond Stove Co., Richmond, VA. $4,500-6,000. *Courtesy of Tiny Moyer.*

296 (not shown). Reno Nevada, 1868: Modern range with black finish, and gold lettering on oven door with "Lake's Crossing" and covered wagon scene. This modern range appears to be produced from the same patterns used by the Grey Iron Casting Co. of Mt. Joy, PA. originally for their largest size "Crescent" range. 13" l., mfgr. possibly The John Wright Foundry, Wrightsville, PA. $25-50.

298. Rival: Ornate, nickel plated range, with two shelves and warming ovens. This range was also produced without the right hand extension shelf. The teakettle shown was only supplied with Stevens' two largest size ranges, the "Eclipse" and the "Rival". 18 1/4" l., mfgr. J&E Stevens. $1,750-2,500.

299. Rival box: Original dovetail wooden box for the J&E Stevens Rival Range. $200-250. *Courtesy of Tiny Moyer.*

300. Rival: Range made in only this one size. Kenton utilized this same design of base for their "Rival" and similar size "Royal." Note that The Dent Hardware Co. has this same design base on their "Queen" pattern range of this size suggesting Dent did pattern work, or produced parts, for Kenton. 8 1/4" l., mfgr. Kenton. $100-150.

301. Roper: Larger of two sizes of Roper gas ranges. This toy gas range was introduced by Arcade Mfg. Co. in 1925, and is a nice replica of the full-size Roper range produced by the George D. Roper Corp. of Rockford, Ill. 5 3/4" l., mfgr. Arcade. $125-175. *Courtesy of Bruce and Melinda Kling.*

303. Roper: Roper range bank No. 844X. Arcade supplied the same range bank in assorted colors, as No, 845X. 3 7/8" l., mfgr. Arcade. $125-175.

302. Roper: Smaller of two sizes of Roper gas ranges. 4 1/4" l., mfgr. Arcade. $100-150.

304 (not shown). Rose: Range of early style. 10 1/2" l., mfgr. uncertain. $500-750.

305. Royal: Twenty different sizes and variations of "Royal" coal/wood ranges were offered by Kenton, by the 1920s. Four examples are photographed in the following listings. Kenton's catalog numbers do vary between their 1905 and 1920s catalogs. Most Kenton ranges were offered with nickel plated or electro oxidized finishes, plus even a nickel/black finish in 1905. Over the years of production the removable backpiece on the larger sizes was changed to include two upper warming ovens with opening doors. Some types were offered with only a stovepipe, and no backpiece.

306. Royal: Largest style Kenton "Royal" range, but without right side extension shelf; electro-oxydized finish. This and other sizes of "Royal" ranges were sold with upper warming ovens in the backpiece in the 1920s. 15 3/4" l., mfgr. Kenton. $400-500. *Courtesy of Tiny Moyer.*

307. Royal: Range with electro oxidized finish and produced with stovepipe only, no backpiece. (No. 931). 11" l., mfgr. Kenton. $300-400.

308. Royal: No. 917 nickel plated range. This design is very similar to the "Pride" range listed, and also the Hubley "Eagle" range of this size. The ranges are similar to Grey Iron's design of the "Cora" range, which serves to illustrate that various firms in the toy industry did produce patterns/parts/complete items for other firms. 6 3/4" l., mfgr. Kenton. $125-175.

310. Royal: Smallest of Kenton's line of "Royal" ranges, complete with correct utensils and backpiece, as issued. No. 903 N. 3 3/4" l., mfgr. Kenton. $75-100.

309. Royal (center), Eagle (left), Pride (right): three similar design ranges, except for the name on the oven door and side shelf. This photo serves to illustrate the toy industry's practice of one firm making patterns, parts or complete items for other toy firms. These three ranges all have a Grey Iron "Cora" range design. The "Royal" is Kenton's No. 917, the "Eagle" is Hubley's 857, and the "Pride" (missing its left side hearth shelf), with an uncertain marketer, are similar. 6 3/4" l., mfgr. Grey Iron or Kenton. $125-175; $125-175; $250-350.

311. Royal: Ornate, nickel plated range. 14" l., mfgr. J&E Stevens. $400-500.

313. Royal: Blue finish range. 8 1/4" l., mfgr. Arcade: $275-375.

312. Royal: Black finish range with gold highlighting. 8 1/4" l., mfgr. Arcade. $275-375.

314. Royal: Gas range with "Broiler." 5 1/4" l., mfgr. A.C. Williams. $75-100.

315. Royal: Single gas burner (left) and double gas burner (right) versions, with aluminum finish. Both sizes are very common. 2 3/4" & 3" l., mfgr. A.C. Williams. $15-20. $20-25.

316. Royal: "Belanger Centenaire 1867-1967," 8" l., mfgr. Belanger, Canada. $25-50. *Photo courtesy of Ralph and Merlyn Wiesehan.*

318. Royal Esther: Nickel plated range, with right side water reservoir. 18" l., mfgr. Mt. Penn Stove Works, Reading. PA. $3,500-4,000.

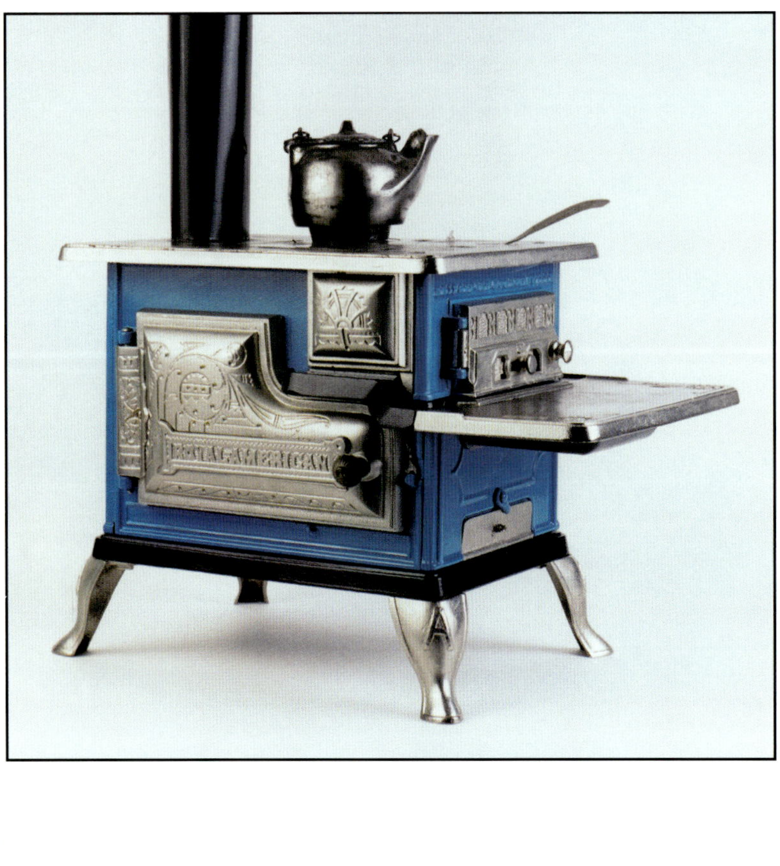

317. Royal American: Stove with blue, black, and nickel plated finish. The letter "A" is on each leg. 18 1/4" l., mfgr. Bridgeford & Co., Louisville, KY. $1,750-2,250. *Courtesy of Tiny Moyer.*

319. Royal Esther: Nickel plated range made without right side water reservoir. 14" l., mfgr. Mt. Penn Stove Works, Reading, PA. $3,000-4,000.

320. Ruby: Small nickel plated range. 7" l., mfgr. J&E Stevens. $300-350. *Courtesy of Bruce and Melinda Kling.*

321 (not shown). Ruth: No details available at time of printing.

S

322. Savage: Yellow painted die-cast and tin gas range. 3 7/8" l., mfgr. Savage, CA. $15-25.

324. Skoglund & Olson: Swedish design range with tiled/brick exterior sides. Skoglund & Olson produced a limited line of cast iron toys in the 1930s, including three ranges. 4 1/4" l., mfgr. Skoglund & Olson, (A-B), Sweden. $850-1,250. *Photo courtesy of Matt Medeiros.*

323. Skoglund & Olson No. 2: Black finish range, shown with original utensils; the larger of two sizes. 14 3/4" l., mfgr. Skoglund et Olson, (A-B), Gefle, Sweden. $1,000-1,500.

325 (not shown). Southern Toy Home: Stove, c. 1899. 15" l., (cooking surface only). Total stove length not ascertained. Mfgr. Southern Stove Works, Evansville, IN. $ N/A.

326. Star: Kenton's second largest range in electro oxidized finish. This range was also produced with a nickel plated finish. 24 1/2" l., mfgr. Kenton. $1,750-2,500. *Courtesy of Tiny Moyer.*

327. Star: Small nickel plated range, with removable left side hearth shelf. 5 3/4" l., mfgr. J&E Stevens. $150-200.

329. Star: Small nickel plated range, with tin sides and simulated stovepipe within removable backpiece. 3 3/4" l., mfgr. Hubley. $ 75-100.

328 (not shown). Star: No. 882S (with utensils). Small nickel plate range, with tin sides and separate stovepipe within removable backpiece. No. 850 was offered with only a rear, centered stovepipe. 3 3/4" l., mfgr. Hubley. $75-100.

330. Star: Small nickel plated range, with tin sides and smaller removable backpiece like the "Tot" range by Grey Iron Casting Co. (No. 893S with utensils). 3 3/4" l., mfgr. Hubley. $50-75.

332. Star: Small nickel plated range, with all cast sides and turnkey underneath to take apart.
3 7/8" l., mfgr. Shimer. $85-125.

331. Star: Small nickel plated range, with all cast sides and turnkey underneath to take apart. 3 1/4" l., mfgr. Shimer. $85-125.

333. Star (five pointed emblem on oven door): Two-burner small range. Similar type to Dent's "W" range. 4 1/2" l., mfgr. Dent. $85-125.

334. Star (five pointed emblem on oven door): Factory pattern for the above listed item. 4 1/2" l., mfgr. Dent. $250-300.

335. St. Clair: Large upright range with black, white porcelain, and nickel finish, and full upper warming oven. This design is quite similar to the large "Midget" range listed, but not photographed, herein. 16" l., mfgr. uncertain (possibly Belleville Stove Works, or Enterprise Foundry Co., both of Belleville, IL.) $4,500-6,000.

336. Sterling: Nickel plated range with two side shelves. The extension shelf is common to other ranges by "Ideal." 13 1/2" l., mfgr. Ideal. $400-500.

337. Superior: Nicely designed nickel plated range, with two shelves and backpiece. The extension shelf design is common to other "Ideal" ranges. As seen in the photo, the "go-with" shown states that full-size "Superior Stoves and Ranges" were produced by "Bridge, Beach, and Co., St. Louis." The backpiece, "lady with bonnet" design in the front base skirt, etc. parallels the Ideal "Princess" range listed herein. 14 7/8" l., mfgr. Ideal. $450-550.

338. Superior: Nickel and black finish range, produced without backpiece or right side extension shelf. The "lady with bonnet" design on oven door and on the front base skirt, and overall characteristics, parallel the Ideal "Princess" range listed herein. A four-burner model of nickel/black "Superior" range was also produced, and is similar in appearance to the "Princess" and "Queen" of the same "series." 12 1/2" l., mfgr. Ideal. $400-500.

338A . Superior: Range with nickel and black finish, with two side shelves and backpiece offset to the left side. This design is similar in appearance to the "Princess" and "Queen" of the same series. 12 1/2 " l., mfgr. Ideal. $400-500.

339. Superior: Nickel plated range with typical Ideal design of the backpiece offset to the left. The oven door design parallels other "Ideal" ranges. 10 1/2" l., mfgr. Ideal. $300-400.

340. Superior: Four sizes of Kenton's "steel gas ranges," which actually burn gas. Patented Nov. 9,1915, except 2nd from left, which was patented prior to 1905. The type shown at the far left was also sold without a side shelf, as No. 994. 9 1/2"-10 3/4"-11 3/4"-6".(L to R), mfgr. Kenton. $750 down to 125. *Courtesy of Tiny Moyer.*

341. Sunshine: Black finish range. Similar to the "Dainty" range of this mfgr. 10 1/2" l., mfgr. Orr, Painter & Co., Reading, PA. $750-850. *Courtesy of Tiny Moyer.*

340A. Superior: Kenton's largest of four sizes of "Steel Gas Ranges," which actually burn gas. Patented Nov. 9,1915. This specimen has an unusual green/yellow paint combination. Normally this range is encountered nickel plated with a blue body. 11 3/4" l., mfgr. Kenton. $750-1,000. *Photo Courtesy of Tiny Moyer.*

342. Sunshine (Design on oven door): Small range, with aluminum finish and unusually thin burner covers (4). 4 3/4" l., mfgr. uncertain. $125-150.

T

343. Tarbox: Range with black and nickel plated finish. This range sits on a separate base. 9 3/4" l., mfgr. Spicers & Peckham, Providence, RI. $800-1,000. *Courtesy of Tiny Moyer.*

344. Thirty–seven (37): Tiny nickel plated range, with turnkey underneath to take-apart. The "36" and "37" are the smallest of Shimer's line. 2 1/2" l., mfgr. Shimer. $125-150.

345. Thirty-six (36): Tiny nickel plated range, with turnkey underneath to take-apart. The "36" and "37" are the smallest of Shimer's line. 2 1/2" l., mfgr. Shimer. $125-150.

346. Tot: Early design nickel plated "Tot" range. 5 1/8" l., mfgr. Grey Iron. $125-150.

348. Tot: Catalog No. 1 1/2" nickel plated small range, equipped with correct stew pan and skillet 3 1/8" l., mfgr. Grey Iron. $50-75.

347. Tot: Somewhat later design nickel plated "Tot" range than the previously listed "Tot." 4 7/8" l., mfgr. Grey Iron. $125-150.

349. Tot: Brass pattern for a size of "Tot" range not noted in production. While this brass pattern is larger than the production "Tot" listed in the Grey Iron Casting Co.'s catalog as No. 1, it was located amongst a group of brass pattern ranges from the Dent Hardware Company. Mfgr. Dent. $250-300.

350. Triumph: Nickel plated version of this range, showing the same backpiece as Kenton's catalog no. 940, in 1905. This range was sold in the 1920s with two warming ovens in the upper backpiece. 15" l., mfgr. Kenton. $700-900. *Courtesy of Tiny Moyer.*

351. Triumph: Electro oxidized version of this range, showing an alternate backpiece design to Kenton's catalog No. 941, in 1905. This range was sold in the 1920s with two warming ovens in the upper backpiece. 15 1/2" l., mfgr. Kenton. $600-700. *Courtesy of Bruce and Melinda Kling.*

352. Two Hundred Two (202): Black enameled range with gold highlighting, left side hearth shelf and stovepipe. The No. "202" on the oven door reflects Wilkins' catalog numbering sequence, which follows their No. 201 "New England Kitchen" range, and No. 200 unnamed "Stove" c. 1890s. 8" l., mfgr. Wilkins. $300-350.

U

353. Uncle Sam Jr.: C. 1876 range with "low hot closet." Another variety was made with left side reservoir. 24 1/4" l., mfgr. Abendroth Bros., Port Chester, N.Y. $2,500-3,500. *Courtesy of Tiny Moyer.*

354. Uncle Sam Jr.: C. 1876 range with reservoir. 23 1/2" l., mfgr. Abendroth Bros., Port Chester, N.Y. $1,500-2,000. *Photo courtesy of Ralph and Merlyn Wiesehan.*

355. Uncle Sam Jr.: C. 1876 range with extension shelf but no reservoir. 23 1/2" l., mfgr. Abendroth Bros., Port Chester, N.Y. $1,250-1,500. *Photo courtesy of Ralph and Merlyn Wiesehan.*

356. Unio: Range with dark finish and stovepipe. Left side hearth shelf is missing. 9" l., mfgr. uncertain. $250-350.

357. Union: Range with copper flash finish and hearth shelf. 8" l., mfgr. Dent. $250-300.

358. Union: Nickel plated range, with hearth shelf. 8" l., mfgr. Dent. $250-300. *Courtesy of Tiny Moyer.*

359. Universal: Largest size Universal Electric Range, in green and nickel with double ovens. Catalog no. 1210. Five sizes of Universal Electric Ranges appear in the 1931 Kenton catalog, but are no longer shown in the 1932 catalog. 9" l., mfgr. Kenton. $750-800.

360. Universal: The two smallest size Universal Electric Ranges, in green and nickel, with single oven. Catalog no. 1200 & 1202. Five sizes of Universal Electric Ranges appear in the 1931 Kenton Catalog, but are no longer shown in the 1932 catalog. 4 1/2" & 6 1/4" l., mfgr. Kenton. $150-200; $300-400.

361. (not shown). Unnamed: Nickel plated range, with two burners, consisting of a series of concentric rings. Size unavailable, mfgr. Barum Vark, Norway. $400-600.

362 (not shown). Unnamed: Stove with sliding front hearth plate, similar in appearance to the "Dictator Jr." and "Cotton Plant" stoves produced in the U.S.A. The firm Carrier & Laine started in business in Levis, Quebec in 1860, on the St. Lawrence River, where it remained for at least ten years, with an output of 2000 stoves per year. Mfgr. Carrier, Laine & Co., Levis, Quebec, Canada. $1,000-1,200.

363. Unnamed: Cast iron range with surrounding towel bar. This range is probably of European origin. 9 1/2" l., mfgr. E H. $400-600.

365. Unnamed: Small nickel finish range. Since this range was produced without a side shelf, and has no name shown, it could be Ideal's " No. O" listed, but not photographed, in their 1894 catalog. 4 1/2" l., mfgr. Ideal. $125-150. *Courtesy of Tiny Moyer.*

364 (not shown). Unnamed: Large, heavy French range with brass trim, and poker which serves also as the lid lifter. 13 1/2" l., mfgr. Forneau de Paris, Odelin. $600-800.

366. Unnamed: Tiny gas range, produced in one length and three colors, blue, green, and ivory, in 1931. Lavender and nickel were earlier finishes. Catalog No. T7. None of Kilgore's catalogs from 1928 to 1931 show the short legged version in the photo, so it must be a later private modification. 2 1/4" l., mfgr. Kilgore. $25-50. *Courtesy of Bruce and Melinda Kling.*

368. Unnamed: Three step stove, with black finish, four legs, and a series of three arched, connected towers on the oven door. 12" l., mfgr. uncertain. $500-700. *Photo Courtesy of Tiny Moyer.*

369. Unnamed: Cooking stove with circular rotating surface, operated by crank handle. 12 1/2" l., mfgr. uncertain. $2,500-3,000. *Courtesy of Tiny Moyer.*

367 (not shown). Unnamed: French style range with water reservoir and front faucet. 12" l., mfgr. St. Nicolas Revin, Canada. $600-750.

370. Unnamed: Small Range with circular, raised vignette of a young lady on the oven door. The scalloped base design on three sides is not common to any other range noted by the authors. 4 5/8" l., mfgr. uncertain. $200-250.

372. Unnamed: No. 200 Range, c. 1895, with cast iron chimney. 9 1/2" l., mfgr. Wilkins. $1200-1500.

371. Unnamed: Lovely range with no markings. The mfgr. of this ornate range was possibly the same as the mfgr. of the "Grand Jewel," perhaps Ideal Mfg. Co., Detroit, MI. 16" l., mfgr. uncertain. $600-800.

373. U.S.: Range with cream and green porcelain finish and two loose trivets. Marked "Gothic 68-18." This range is similar in design to the "Mascot," "Wetter," and "Hanks Junior," which is indicative of the progression of stove/range manufacturers utilizing the same basic patterns over a span of many years. 18 1/2" l., mfgr. U.S. Stove Corp., South Pittsburg, TN. $3,500-4,000. *Courtesy of Tiny Moyer.*

V

374. Venus: Range with black finish and "star and crescent" motif on oven door. The Shimer catalog also illustrates a "Venus" range with different motif, at 6 1/2" length. 7 3/4" l., mfgr. Shimer. $400-450.

375. Venus: Range with nickel plated and black finish. 8 1/2" l., mfgr. Kenton. $175-200.

W

376. W: Small nickel plated range with the initial "W" on the oven door. *Note:* see similar small range with five-pointed "star" on oven door. 4 1/2" l., mfgr. Dent. $85-125

378. Waldorf: Small range with black and nickel finish and stovepipe. See the brass pattern also listed for this four-burner range. 4 7/8" l., mfgr. Dent. $125-150.

377. Waldorf: Brass pattern for a four-burner version of the "Waldorf" range with two shelves and backpiece. 10 3/4" l., mfgr. Dent. $500-750.

379. Waldorf: Brass pattern for this small range, with stovepipe. See the production issue also listed for this four-burner range, which is 1/8" smaller due to casting shrinkage. 5" l., mfgr. Dent. $300-400.

380. Waldorf: Brass pattern for two-burner version of this small range, with stovepipe. 4 3/4" l., mfgr. Dent. $250-300.

381. Walkers: Gas range with private label. Similar to Arcade's "Roper" range, but doors and hinges are similar to their larger "Peninsular" range. 5 1/2" l., mfgr. Arcade. $300-450. *Photo courtesy of Donna Johnson.*

382. Wedgewood: Large range with porcelain and nickel finish. Similar to the Buck's Junior 2 range listed. The "Wedgewood" name was used on full size ranges produced by James Graham Mfg. Co. 21" l., mfgr. uncertain, perhaps Buck's Stove & Range Co. for Graham Mfg. Co. $2,800-3,900. *Courtesy of Tiny Moyer.*

383 (not shown). West: Modern die-cast home decorator range with copper flash finish and non-removable burner lids. 9" l., mfgr. Stern Industries, Inc.$15-25.

384. Wetter: Range with blue porcelain finish, and two loose trivets, called "U.S. Baby Gothic" by mfgr. Produced c. 1930. See similar design ranges named: "U.S.." "Mascot," "Baby Hazel," and "Hanks Junior."18" l., mfgr. U. S. Stove Corp., South Pittsburg, Tenn., formerly H. Wetter Mfg. Co./ Blue City Stove Works, Memphis, Tenn. $3,000-3,500. *Photo courtesy of Ralph and Merlyn Wiesehan.*

386. Wincroft: Cream and nickel plated gas range. Similar to the "Roper" range made by Arcade Mfg. Co. (shown with other "Wincroft" advertising "go-withs"). 5 1/2" l., mfgr. Arcade. $300-450.

385. Williams: Small green painted gas range, with red base and silver color cooking surface. c. 1936. Shown with complementing kitchen sink unit sold concurrently by A.C. Williams. 3 5/8" l., (range), 4 1/8" l., (sink), mfgr. A.C. Williams. $75-100.

XYZ

387. XX: Small four-burner range, sold both with a plain finish and nickel plated. Sold both with and without utensils. 4 7/8" l., mfgr. Shimer. $100-125.

389. Young, Peerless, Buck & Wright: Cook stove, pat. 1860. 8 3/4" l., mfgr. Young, Peerless, Buck & Wright-St. Louis. $800-1,000. *Photo courtesy of Ralph and Merlyn Wiesehan.*

388. York: Gas range bank accompanied by a contemporary pinback button, showing a lady cooking on a full-size "York" gas range of this type, with the slogan, "Use a York Gas Range." 4" l., mfgr. Abendroth Bros., N.Y. $250-350.

390. Zenda: Large nickel plated range, c. 1900. 17 3/4" l., mfgr. Kenton. $600-800.

392. Zoo: Small range made with stovepipe only, no shelves or backpiece. Two designs of base were used on the "Zoo" range, with and without legs. The "1900" range is the same except for the oven door. 4 1/8" l., mfgr. Kenton. $100-125.

391 (not shown). Zenith: Large nickel plated range with two removable shelves. The opening front doors mount on pins similar to other large ranges c. 1900 by Kenton. Approx. 16" l., mfgr. probably Kenton. $600-800.

Index

This alphabetical index serves as a cross-reference to the images presented in Chapters Five, Six, and Seven and the numerical listings in Chapter Eight. Items with a "GW" prefix are located in Chapter Five, "Additional 'Go-Withs' For Toy Stoves and Ranges," and those with a "CF" prefix are located in Chapter Six, "Complementary Fields of Collecting." Those items with a prefix of ID are found in Chapter Seven "Identification and Variations."